Solve It!
Problem-Solving Strategies

Research and Development
Betty Cordel
Judith Hillen
Myrna Mitchell
Michelle Pauls
Jim Wilson
Dave Youngs

Illustrators
Dawn McAndrews
Brock Heasley
Ben Hernandez
Renée Mason
Margo Pocock
David Schlotterback
Brenda Wood

Desktop Publishers
Tracey Lieder
Roxanne Williams

Solve It! 4th:
Problem-Solving Strategies

Developed and Published by

AIMS Education Foundation

This book contains materials developed by the AIMS Education Foundation. **AIMS** (**A**ctivities **I**ntegrating **M**athematics and **S**cience) began in 1981 with a grant from the National Science Foundation. The non-profit AIMS Education Foundation publishes hands-on instructional materials that build conceptual understanding. The foundation also sponsors a national program of professional development through which educators may gain expertise in teaching math and science.

Copyright © 2006, 2009, 2012 by the AIMS Education Foundation

All rights reserved. No part of this book or associated digital media may be reproduced or transmitted in any form or by any means—except as noted below.

- A person or school purchasing this AIMS publication is hereby granted permission to make up to 200 copies of any portion of it (or the files on the accompanying disc), provided these copies will be used for educational purposes and only at one school site.

- For a workshop or conference session, presenters may make one copy of any portion of a purchased activity for each participant, with a limit of five activities or up to one-third of a book, whichever is less.

- All copies must bear the AIMS Education Foundation copyright information.

- Modifications to AIMS pages (e.g., separating page elements for use on an interactive white board) are permitted only within the classroom or school for which they were purchased, or by presenters at conferences or workshops. Interactive white board files may not be uploaded to any third-party website or otherwise distributed. AIMS artwork and content may not be used on non-AIMS materials.

AIMS users may purchase unlimited duplication rights for making more than 200 copies, for use at more than one school site, or for use on the Internet. Contact us or visit the AIMS website for complete details.

AIMS Education Foundation
1595 S. Chuestnut Ave., Fresno, CA 93702-4706
888.733.2467 • aimsedu.org

ISBN 978-1-60519-082-2

Printed in the United States of America

Solve It! 4th:
Problem-Solving Strategies

Chinese Proverb 4	*Look for Patterns* Introduction 163
Introduction 5	Fabulous Four-Sum 165
Problem-Solving Strategies 6	Crack the Code 177
Strategy Reference 7	Code Consultant 183
	Tallying the Times Table 189
Use Manipulatives Introduction 9	
Twenty-4 Square 11	*Use Logical Thinking* Introduction 195
Select-a-Square 17	Made-to-Order Rectangles 197
Fiddler Fun 23	Shape Logic 207
	Nabbing Numbers 215
Write a Number Sentence Introduction .. 37	
Addition Rummy 39	*Wish for an Easier Problem* Introduction ... 223
To Add, or Not to Add? 47	Our Library in the Limelight 225
Parks Postage Problem 57	Calculating a Classroom Crunch 231
	A Close Call 239
Draw out the Problem Introduction 65	
Key Counts 67	Practice Problems 247
Finding Floors and Reckoning Rungs ... 71	
Polar Passage 77	
Work Backwards Introduction 83	
Primarily Pro-bear-bility 85	
Fuel Filling Figures 93	
Mix-Ups and Mysteries 101	
Organize the Information Introduction .. 111	
Dots by the Dozen 113	
That's Sum Pattern! 121	
Patty's Penny Puzzler 127	
Centennial Celebration Calculations 133	
Guess and Check Introduction 139	
Square One 141	
Time Pieces 147	
X-cellent Addition 151	

I Hear and I Forget,

I See and I Remember,

I Do and I Understand.

-Chinese Proverb

Introduction

Solve It! 4th: Problem-Solving Strategies is a collection of activities designed to introduce students to nine problem-solving strategies. The tasks included will engage students inactive, hands-on investigations that allow them to apply their number, computation, geometry, data organization, and algebra skills in problem-solving settings.

It can be difficult for teachers to shift from teaching math facts and procedures to teaching with an emphasis on mathematical processes and thinking skills. One might ask why problem solving should be taught at all. The most obvious reason is that it is part of the mathematics curricula. Indeed, with the introduction of the Common Core State Standards for Mathematics, there has been a renewed emphasis on the so-called standards for mathematical practice, the first of which is problem solving ("Make sense of problems and persevere in solving them."). Placing an emphasis on the standards for mathematical practice (as is recommended by the Common Core) is an interesting and enjoyable way to extend and deepen the learning of mathematics; it encourages collaborative learning, and it is a great way for students to practice the application of mathematical skills. This in turn leads to better conceptual understanding—an understanding that allows students to remember and to apply these skills in different contexts.

Introducing students to the nine strategies included in this book gives them a toolbox of problem-solving methods that they can draw from when approaching problems. Different students might approach the same problem in a variety of ways, some more sophisticated than others. Hopefully, every child can find one approach that he or she can use to solve the problems that you present. Over time, and from discussing what other children have done, students will develop and extend the range of strategies at their disposal.

It is our hope that you will use the problems in this book to enrich your classroom environment by allowing your students to truly experience problem solving. This means resisting the urge to give answers; allowing your students to struggle, and even be frustrated; focusing on the process rather than the product; and providing multiple, repeated opportunities to practice different strategies. Doing this can develop a classroom full of confident problem solvers well equipped to solve problems, both in and out of mathematics, for years to come.

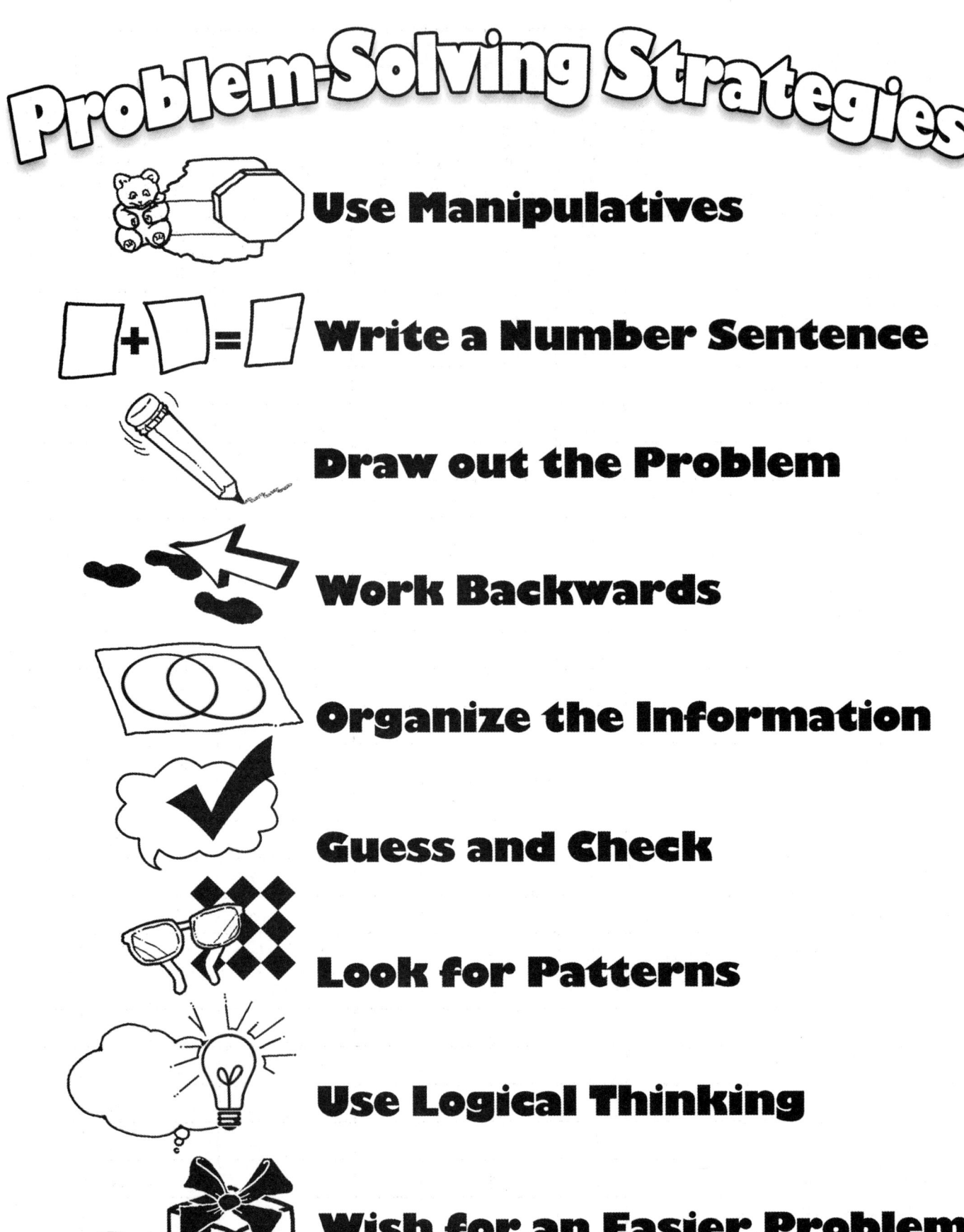

Activities \ Strategies	Use Manipulatives	Write a Number Sentence	Draw out the Problem	Work Backwards	Organize the Information	Guess and Check	Look for Patterns	Use Logical Thinking	Wish for an Easier Problem
Twenty-4 Square	X						X		
Select-a-Square	X				X		X		
Fiddler Fun	X						X		
Addition Rummy	X	X							
To Add, or Not to Add?		X					X		
Parks Postage Problem	X	X			X		X		
Key Counts			X		X		X		
Finding Floors and Reckoning Rungs			X						
Polar Passage	X		X					X	
Primarily Pro-bear-bility				X		X			
Fuel Filling Figures				X		X			
Mix-Ups and Mysteries	X			X					
Dots by the Dozen	X				X		X		
That's Sum Pattern!					X	X	X		
Patty's Penny Puzzler					X	X	X		
Centennial Celebration Calculations			X		X				
Square One	X						X	X	
Time Pieces	X						X		
X-cellent Addition	X						X	X	
The Fabulous Four-Sum	X				X		X		
Crack the Code							X		
Code Consultant							X		
Tallying the Times Table					X		X		
Made-to-Order Rectangles	X					X		X	
Shape Logic	X					X		X	
Nabbing Numbers	X							X	
Our Library in the Limelight					X				X
Calculating a Classroom Crunch					X				X
A Close Call				X	X				X

Problem-Solving Strategies
Use Manipulatives

Sometimes it is helpful to use objects when solving a problem. These objects can represent the parts of the problem. Seeing the parts can help you understand how to find the answer. Anything can be a manipulative. You can use paper clips, pattern blocks, Unifix cubes, or even pieces of paper.

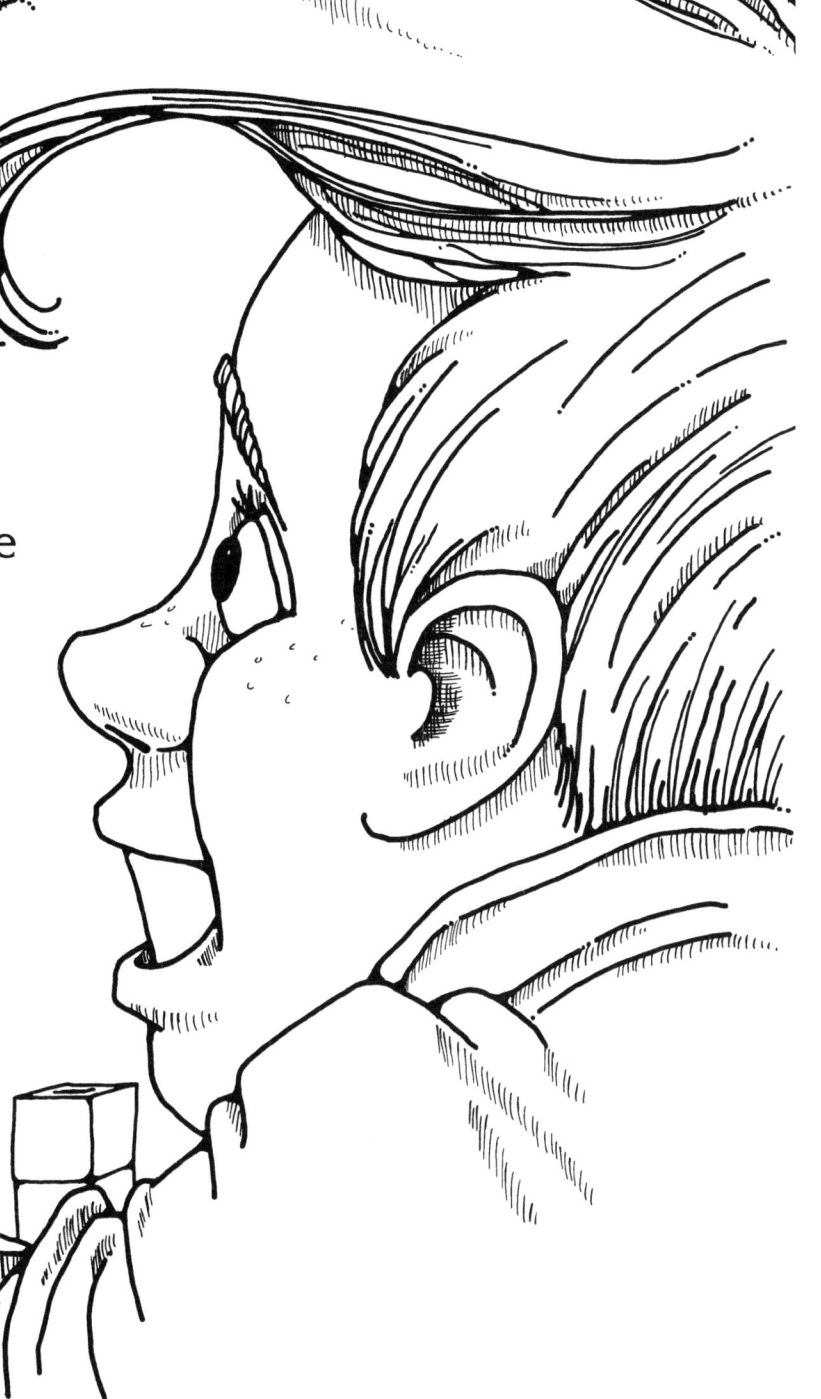

TWENTY-4 SQUARE

Topic
Area and perimeter

Key Question
How many different rectangles can be made using exactly 24 squares?

Learning Goals
Students will:
- use 24 squares to make all possible rectangles,
- explore how the perimeter of these rectangles changes, and
- think of related extensions to explore.

Guiding Documents
Project 2061 Benchmark
- Length can be thought of as unit lengths joined together, area as a collection of unit squares, and volume as a set of unit cubes.

Common Core State Standards for Mathematics*
- Make sense of problems and persevere in solving them. (MP1)
- Construct viable arguments and critique the reasoning of others. (MP3)
- Understand concepts of area and relate area to multiplication and to addition. (3.MD)
- Recognize perimeter as an attribute of plane figures and distinguish between linear and area measures. (3.MD)

Math
Geometry
  rectangles
  area and perimeter
Number and operations
  multiplication
  factors
Problem solving

Integrated Processes
Observing
Comparing and contrasting
Collecting and recording data
Interpreting data

Problem-Solving Strategies
Use manipulatives
Look for patterns

Materials
Area Tiles, 24 per student
Student pages

Background Information
Area and perimeter are two key mathematical concepts that are an integral part of this activity. Since students deal with these concepts concretely, they have a better chance to understand them. Students should discover that while all the rectangles have areas of 24 square units, their perimeters change. They will discover that the 1 x 24 and the 24 x 1 rectangles have the greatest perimeters and that the 4 x 6 and 6 x 4 rectangles have the smallest perimeters. The other rectangles have perimeters somewhere in between.

Two other math concepts imbedded in this activity are multiplication and factors. While these concepts are not the focus of the activity, they can be used to enrich it. Students can be shown that each rectangle constructed is a concrete representation of a multiplication problem. For example, the 3 x 8 rectangle (three rows of eight squares each) is an array model for 3 x 8 = 24. In the same way, students can be shown that the sides of the rectangles (three and eight in the example) are factors of 24.

Management
1. It is important that students have squares to manipulate when doing this activity. Area Tiles work well and can be purchased from AIMS. If these, or other square manipulatives, are not available, a page with two sets of 24 squares has been provided. Run this off on card stock and have students cut the squares apart.
2. Students should work together in pairs or small groups.

Procedure
1. Distribute one set of 24 squares to each student along with the student page.
2. Go over the challenge and provide time for groups to discover all of the solutions. Tell students that they should include all rectangles, even if some have the same dimensions in a different order.
3. Have groups share their solutions and some of the discoveries they made.
4. Encourage students to explore some of the extensions they thought of and to report their findings.

Connecting Learning
1. How many rectangles can you make with 24 squares? [eight]
2. How do you know that you have found them all?
3. How are the dimensions of the rectangles related to multiplication? [The dimensions represent the whole-number factors of 24.]

SOLVE IT! 4th © 2012 AIMS Education Foundation

4. Do all of the rectangles represent different multiplication problems? Justify your response. [While the answer is the same for both rectangles, the arrays and the multiplication problems they model are different. It is not correct to say that 6 x 4 and 4 x 6 are equal. The first means six sets of four, while the second means four sets of six.]
5. What happens to the perimeters of the rectangles as the lengths of the sides change?
6. What other extensions would you like to explore?

Extensions
1. Have students explore how many rectangles can be made from a different number of squares. If 23 squares are used, students will discover that only two rectangles (1 x 23 and 23 x 1) can be made since 23 is prime and has only two factors. If 25 squares are used, only three rectangles can be made, 1 x 25, 5 x 5, and 25 x 1. Students should be able to see that while 25 is not prime, it does not have as many factors as 24 and therefore fewer rectangles are possible. If the number of squares used is another number with more factors, such as 36 or 48, students will find that many rectangles are possible.
2. Explore the non-rectangular shapes that can be made with 24 squares. What is the maximum perimeter possible?

Solutions
There are eight possible rectangles that can be made using 24 squares.

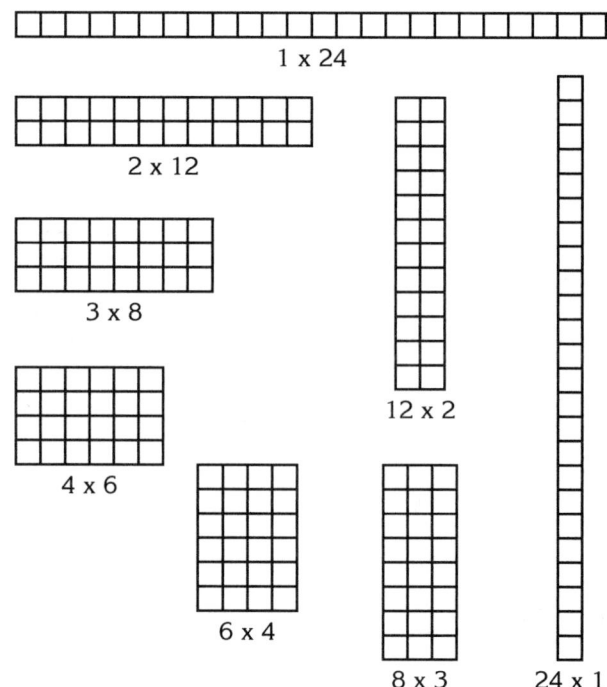

* © Copyright 2010. National Governors Association Center for Best Practices and Council of Chief State School Officers. All rights reserved.

TWENTY-4 SQUARE

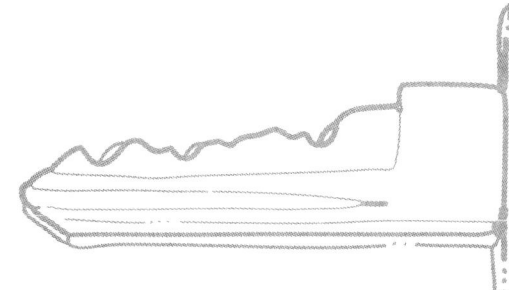

Key Question

How many different rectangles can be made using exactly 24 squares?

Learning Goals

Students will:

- use 24 squares to make all possible rectangles,

- explore how the perimeter of these rectangles changes, and

- think of related extensions to explore.

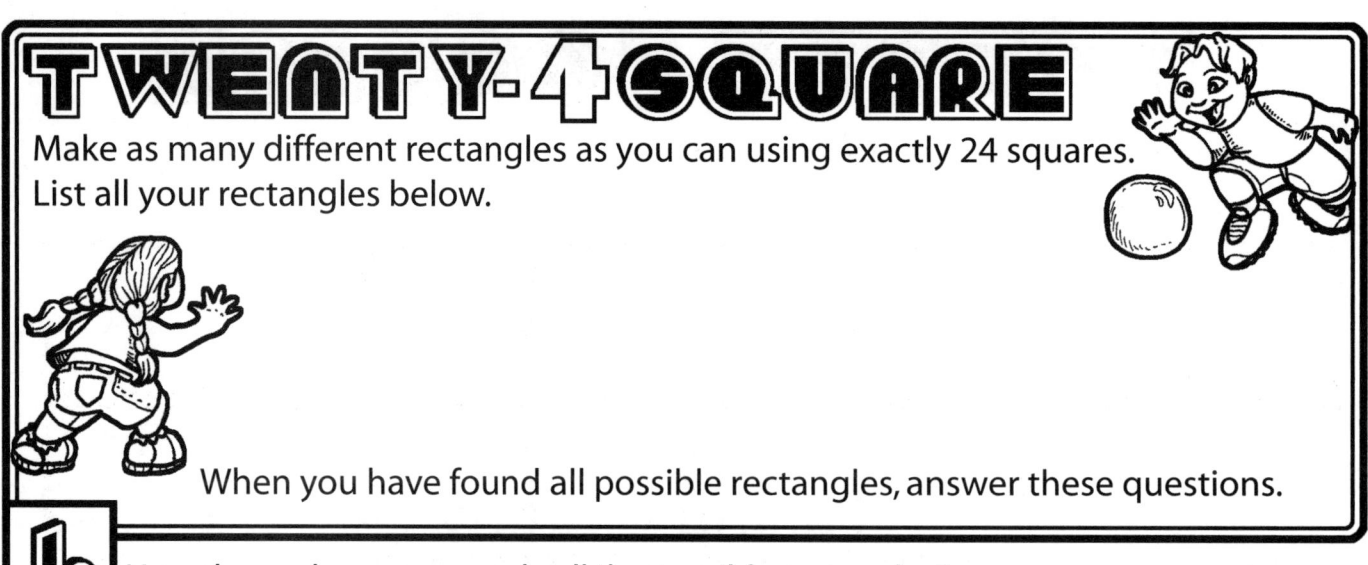

TWENTY-4 SQUARE

Make as many different rectangles as you can using exactly 24 squares. List all your rectangles below.

When you have found all possible rectangles, answer these questions.

 How do you know you made all the possible rectangles?

 Using words, pictures, charts, etc., describe the process you used to solve this problem.

 How are the dimensions of the rectangles related to multiplication?

 Find the perimeter of each of your rectangles. Describe what you discover. Why do you think this happens?

 Write some additional extensions you would like to explore on the back of this paper.

SOLVE IT! 4th   14   © 2012 AIMS Education Foundation

TWENTY-4 SQUARE

Cut apart one set of 24 squares. Make as many different rectangles as you can using exactly 24 squares.

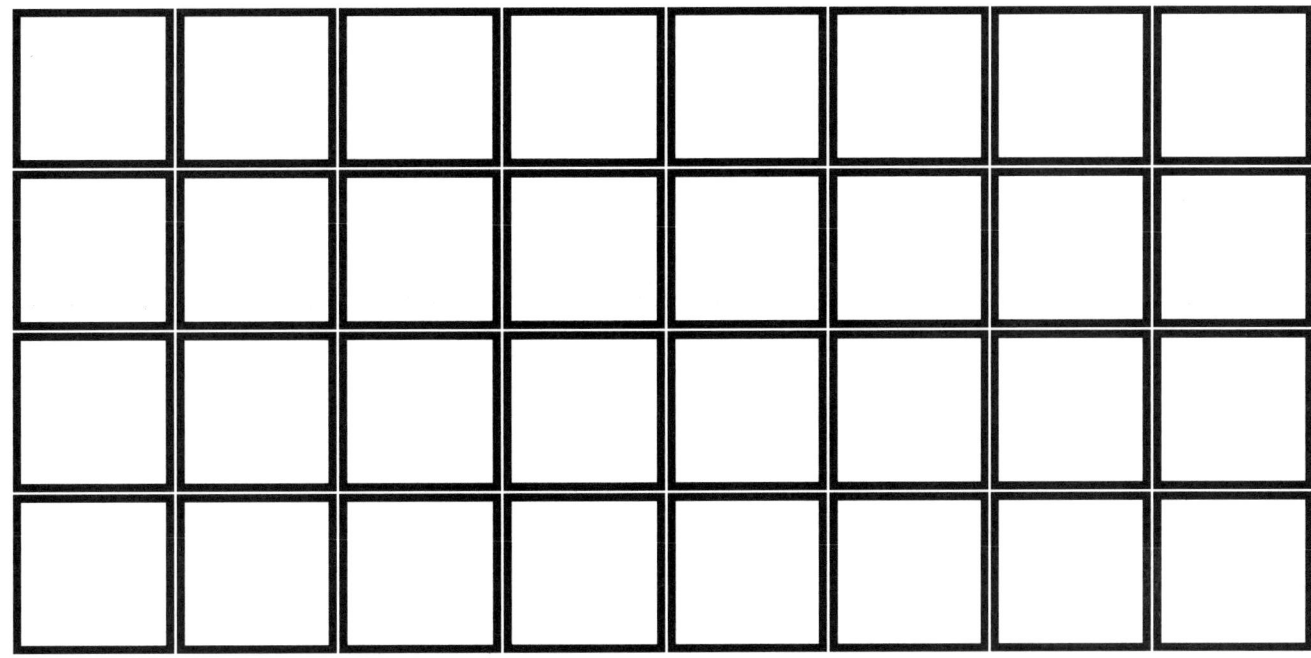

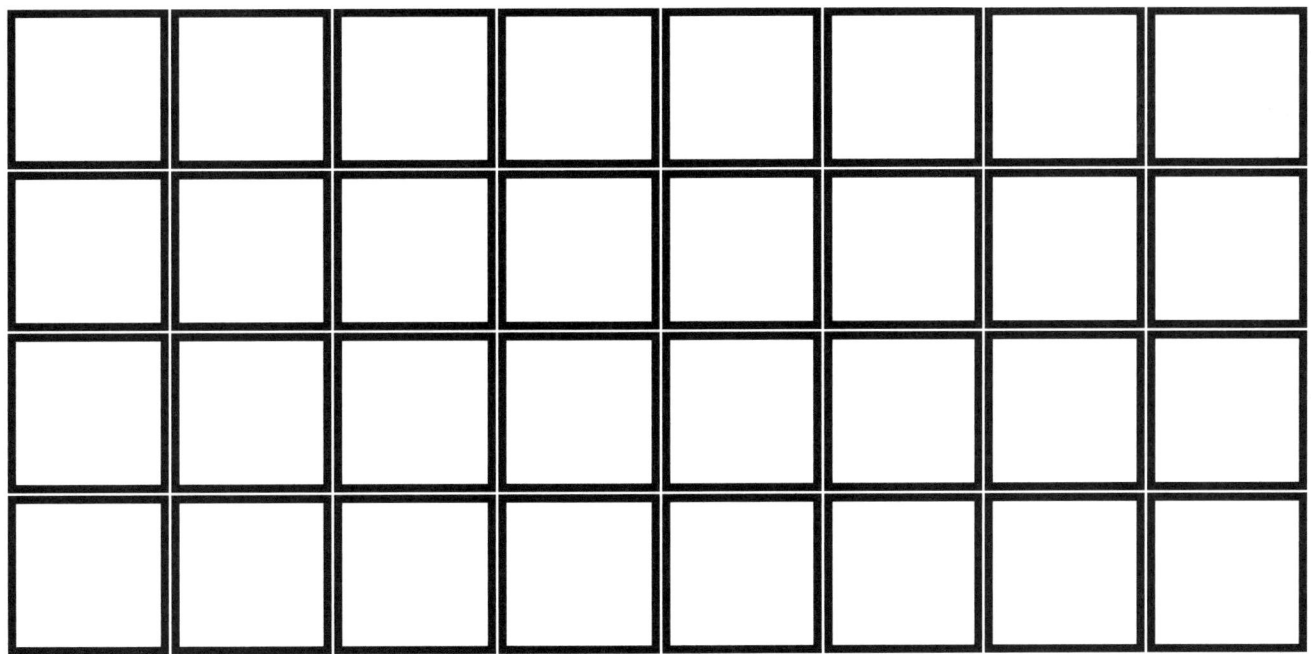

TWENTY-4 SQUARE

Connecting Learning

1. How many rectangles can you make with 24 squares?

2. How do you know that you have found them all?

3. How are the dimensions of the rectangles related to multiplication?

4. Do all of the rectangles represent different multiplication problems? Justify your response.

5. What happens to the perimeters of the rectangles as the lengths of the sides change?

6. What other extensions would you like to explore?

SELECT·A·SQUARE

Topic
Problem solving

Key Question
How can you select squares out of grids so that none of the selected squares are in the same row and column?

Learning Goals
Students will:
- find all of the ways to select three squares from a 3 x 3 grid so that none of the selected squares are in the same row or column,
- record and organize their solutions based on patterns, and
- try to predict the number of solutions for the same problem using a 4 x 4 grid.

Guiding Document
Common Core State Standards for Mathematics*
- *Make sense of problems and persevere in solving them. (MP1)*

Math
Problem solving

Integrated Processes
Observing
Comparing and contrasting
Recording
Predicting

Problem-Solving Strategies
Use manipulatives
Look for patterns
Organize the information

Materials
Area Tiles, 3 per student
Student pages

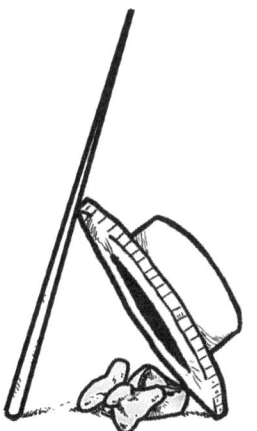

Background Information
In *Select-a-Square*, students are presented with square grids of differing sizes—2 x 2 and 3 x 3. They are challenged to select two squares from the 2 x 2 grid and three squares from the 3 x 3 grid so that none of the squares are in the same row or column. They are then asked to identify the number of solutions for each level and justify why they believe they have found them all. Students are also challenged to look for patterns, to organize their solutions based on these patterns, and to try to predict the number of solutions for 4 x 4 grid. This simple activity provides the opportunity for practicing a variety of problem-solving strategies, including use manipulatives, look for patterns, and organize the information.

Management
1. Each student will need three Area Tiles (available from AIMS) or a similar small manipulative that will fit in the squares on the student page.
2. Make a copy of the both student pages for each student.

Procedure
1. Distribute the first student page and Area Tiles to students. Go over the instructions to make sure everyone is clear on the task.
2. Provide time for students to work on the problem and come up with their solutions.
3. Once students have finished, distribute the second student page and give them time to answer the questions.
4. Close with a time of discussion where students share their solutions, the patterns they found, and the methods they used for organizing their solutions.

Connecting Learning
1. How many solutions are there for the 2 x 2 square? How do you know you have found them all?
2. How many solutions are there for the 3 x 3 square? How do you know you have found them all?
3. What patterns did you discover in your solutions?
4. How did you use these patterns to organize your solutions?
5. Was the way you organized your solutions the same as or different from your classmates? Why?
6. Do you think one method is better than the others? Why or why not?
7. How many solutions do you think there are for a 4 x 4 square? Why do you think this?

SOLVE IT! 4th   17   © 2012 AIMS Education Foundation

Extension
Challenge students to discover and record the more than 20 solutions for the 4 x 4 grid.

Solutions

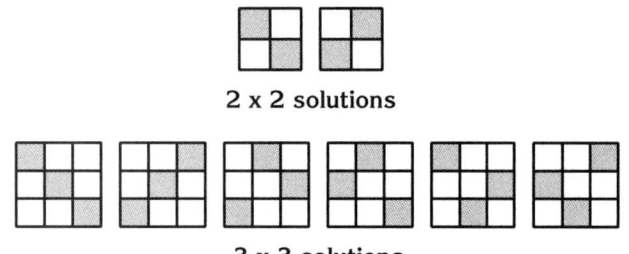

As the teacher, it is important for you to have a good understanding of the patterns, even if your students can only grasp them at the most basic level. As you can see, the number of solutions increases very quickly from one level to the next. While there are only two possible solutions for a 2 x 2 grid, that number triples to six possible solutions for a 3 x 3 grid, and there are more than 20 solutions for a 4 x 4 grid. (The pattern that governs the number of solutions for a given level is fairly advanced, and should not be addressed unless students are very proficient at pattern discovery. The number of solutions for an $n \times n$ grid is $n!$ [read "n factorial"]. $n!$ is the product of all of the positive whole numbers between 1 and n.)

The fourth question on the second student page asks students to describe three different ways in which they could organize their solutions based on the visual patterns that they discover. Several different methods of organization that your students might use are shown here.

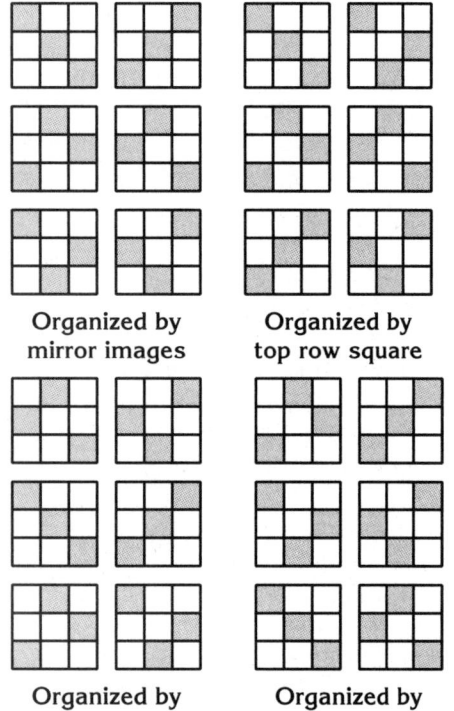

Each of these methods provides a check to help students determine if they have discovered all of the solutions. As you can see in the first set of solutions, each grid has a mirror image. Once this is discovered, students can quickly generalize that the total number of solutions must be even, and that each solution should have a pair. This allows missing solutions to be easily identified and recorded.

Another pattern that can be observed involves the number of times a given square is selected. For example, in the 3 x 3 grids, each of the squares is selected in a total of two solutions. The last three sets of solutions are organized based on this pattern. Again, when this pattern is observed, it provides a simple way for students to fill in any missing solutions.

* © Copyright 2010. National Governors Association Center for Best Practices and Council of Chief State School Officers. All rights reserved.

SELECT·A·SQUARE

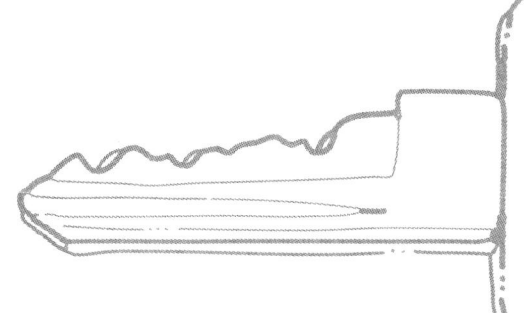

Key Question
How can you select squares out of grids so that none of the selected squares are in the same row and column?

Learning Goals

Students will:

- find all of the ways to select three squares from a 3 x 3 grid so that none of the selected squares are in the same row or column,

- record and organize their solutions based on patterns, and

- try to predict the number of solutions for the same problem using a 4 x 4 grid.

Your challenge is to find every way that you can select squares from within each of the grids below so that no two squares are in the same row or column. You must select two squares for the 2 x 2 grid and three squares for the 3 x 3 grid. Move your marking chips around in the large grids to help you discover all of the solutions. Once you find a solution, record it in one of the smaller grids. One solution for each level has been done to get you started. (You may not need every grid.)

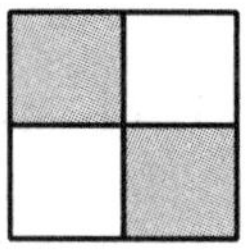

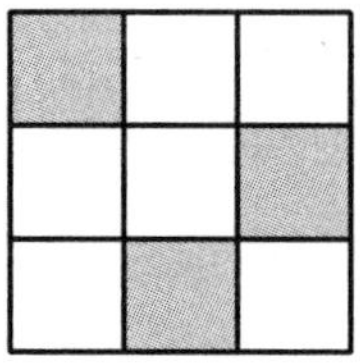

SOLVE IT! 4th 20 © 2012 AIMS Education Foundation

Study your solutions and answer the following questions.

1. How many solutions are there for a 2 x 2 grid? How do you know you have found them all?

2. How many solutions are there for a 3 x 3 grid? How do you know you have found them all?

3. Describe any patterns you see in the way the solutions look. How can these patterns help you determine if you have found all of the solutions?

4. Describe three different ways you could organize your solutions based on the patterns you discovered. Use the grids provided to show one of these ways.

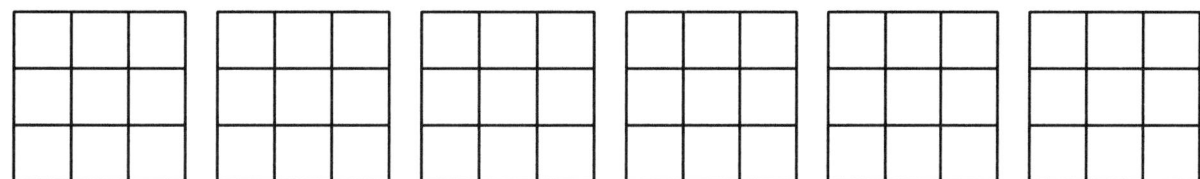

5. How many solutions do you think there would be for a 4 x 4 grid? Why?

Connecting Learning

1. How many solutions are there for the 2 x 2 square? How do you know you have found them all?

2. How many solutions are there for the 3 x 3 square? How do you know you have found them all?

3. What patterns did you discover in your solutions?

4. How did you use these patterns to organize your solutions?

5. Was the way you organized your solutions the same as or different from your classmates? Why?

6. Do you think one method is better than the others? Why or why not?

7. How many solutions do you think there are for a 4 x 4 square? Why do you think this?

Fiddler Fun

Topic
Number sequences and series

Key Questions
1. What geometric patterns do the natural (counting) and odd numbers make?
2. What numerical series are formed from the natural and odd number sequences?

Learning Goal
Students will explore the geometric patterns and numerical series formed by the natural (counting) and odd numbers.

Guiding Documents
Project 2061 Benchmark
- Mathematics is the study of many kinds of patterns, including numbers and shapes and operations on them. Sometimes patterns are studied because they help to explain how the world works or how to solve practical problems, sometimes because they are interesting in themselves.

Common Core State Standards for Mathematics*
- Make sense of problems and persevere in solving them. (MP1)
- Look for and make use of structure. (MP7)
- Generate and analyze patterns. (4.OA)

Math
Number sense
    figurate numbers
    odd and even
Number and operations
    addition
    multiplication
Number patterns
Problem solving

Integrated Processes
Observing
Comparing and contrasting
Recording
Interpreting
Generalizing

Problem-Solving Strategies
Use manipulatives
Look for patterns

Materials
Area Tiles or pennies, 55 per student
Transparent tape
Scissors
Self-adhesive dots, 55 per student
Transparency film
Student pages

Background Information
Fiddlers are an easy-to-make math manipulative that help students develop number sense. A Fiddler is a sequence of one or more square or round tiles taped together. Fiddlers lend themselves to exploring the numbers they represent. The flexible unit construction makes it easy to model whether a given Fiddler is odd or even, what its divisors are, and the family of addition facts that sum to the number on the Fiddler.

In this activity, students use Fiddlers to explore triangular and square number sequences. A *number sequence* is a list of numbers formed by a rule. For example, the counting numbers are made by starting with the number one with the rule that the next number in the sequence is made by adding one to the previous number in the sequence. A *series* is the sum of the terms of a sequence.

In this activity, the sequence of natural numbers (1, 2, 3, 4, 5, ...) forms the series 1, 3, 6, 10, ... called the triangular numbers because each number can be represented geometrically as a triangle. The nth triangular number can be found by using the formula $(n + 1)(n/2)$. For example, the fifth triangular number can be computed by substituting 5 for n in the formula. This gives $(5 + 1)(5/2) = (6)(5/2) = (3)(5) = 15$.

The sequence of odd numbers (1, 3, 5, 7, 9, ...) forms the series 1, 4, 9, 16, 25, ... called the square numbers because each number can be represented geometrically as a square. The nth square number can be computed by using the formula n^2. For example, the 6th square number is $6 \times 6 = 36$.

Management
1. Organize students into groups of two. Each student should have a set of Fiddlers that is a different color than the other student in the group.
2. The parts of this activity can be done on different days if desired.
3. Make transparencies of the *Fiddler Grid* and *Oddly Square* pages.

SOLVE IT! 4th   © 2012 AIMS Education Foundation

Procedure

Part One—Introduction
1. Distribute the materials necessary for each student to make a set of Fiddlers. Pair the students so they can help each other build their sets.
2. Give each pair of students the *Counting Fiddlers* page. Allow time for some free exploration with the Fiddlers as students answer the questions.
3. Discuss the answers to the questions and any discoveries that students made while working with their Fiddlers.

Part Two—Triangular Numbers
1. Distribute one *Fiddler Grid* page to each group.
2. Use the *Fiddler Grid* transparency and a set of Fiddlers to show the students how to place the Fiddlers on the grid according to the instructions.
3. Instruct the students to take turns placing Fiddlers on the grid and recording the numbers along the left side of the grid.
4. Distribute one copy of the *Number Patterns* page to each student.
5. Have the students record within the braces ({ _ }) the number sequence that appears along the left side of the grid.
6. Ask the students to record the *rule* that forms the sequence. (See *Background Information*.)
7. Have the students write the name of the sequence.
8. Tell the students to remove all but the 1- and 2-Fiddler from the grid and observe that a triangle can be drawn that passes through the three perimeter (outside) points. Have them add a 3-Fiddler to the grid and then draw a triangle through the perimeter points of the drawing. Have the students repeat this process for the patterns formed by adding a 4- and then a 5-Fiddler to the grid.
9. Ask the students to come up with a name for the sequence of numbers that lie along the diagonal of the grid. [triangular numbers]

Part Three—Square Numbers
1. Repeat the procedures from *Part Two* using the *Oddly Square* and *Oddly Square Patterns* pages.
2. Discuss student responses and how the square numbers and the triangular numbers compare.
3. Encourage students to think of other ways they can use their Fiddlers.

Connecting Learning

Part One—Introduction
1. What things did you discover about your Fiddlers?
2. What kinds of problems can you solve with your Fiddlers?
3. What do your Fiddlers tell you about the numbers they represent?

Part Two—Triangular Numbers
1. What are the triangular numbers? [1, 3, 6, 10, ...]
2. How are they made? [by adding the consecutive natural (counting) numbers]
3. Why do you think they are called triangular numbers? [They can be represented geometrically by triangles.]
4. What are the next two numbers in the series? [45 and 55] Are these two numbers also triangular numbers? Explain. [yes]
5. Arrange all 10 Fiddlers on the grid page. How does the number of tiles down the left side, along the bottom, and along the diagonal compare? [They are the same.]
6. If you had an 11-Fiddler to add to the bottom, would your answer to the comparison question change? Explain. [No. Adding the 11-Fiddler adds one tile to the left side, one tile to the bottom, and one tile to the diagonal.]
7. Using both sets of Fiddlers, what's the largest triangular number you can build? [105]
8. If you were to continue adding larger Fiddlers in counting order, will each successive number in the series be a triangular number? Explain. [Answers will vary, but one possible explanation could be that if the last Fiddler added to the bottom row of a triangular number was n, then the next Fiddler, in order, would be $n + 1$. But the $n + 1$ Fiddler would add one to the left side of the triangle and one to the diagonal so all three sides would be $n + 1$ which means the pattern represents the $(n + 1)$ triangular number.]

Part Three—Square Numbers
1. What are the square numbers? [1, 4, 9, 16, ...]
2. How are they made? [by doubling natural (counting) numbers or by adding consecutive odd numbers]
3. Why do you think they are called square numbers? [They can be represented geometrically by squares.]
4. Using both sets of Fiddlers, what's the largest square number you can build? [100]
5. If the grid were as large as you wanted and you were to continue adding two consecutive Fiddlers in counting order, would each successive number in the series be a square number? Explain. [Answers will vary, but one possible explanation could be that if the last Fiddler added to the bottom row of a triangular number was n, then the other consecutive Fiddler placed vertically would be $n - 1$. But, since the lower right tile is counted as both a row and a column tile, the number of tiles has to $n \times n$, which is a square number.]

Solutions

Counting Fiddlers
- The groups of 10 that can be formed using two Fiddlers are 1 + 9, 2 + 8, 3 + 7, 4 + 6, and 5 + 5.
- The sum of the groups of 10 is 50.
- The 10-Fiddler wasn't used to form a group of 10 because it's already worth 10.
- The sum of all the Fiddlers is 55.

Fiddler Grid

	1							
1	1							
2		3						
3			6					
4				10				
5					15			
6						21		
7							28	
8								36

Number Patterns
- The sequence along the left side of the *Fiddler Grid* is {1, 2, 3, 4, 5, 6, 7, 8}.
- The sequence, called the natural (counting) numbers, is made by adding one to the previous number.
- The series of numbers that lies along the diagonal of the grid is {1, 3, 6, 10, 15, 21, 28, 36}.
- This series is referred to by mathematicians as the *triangular numbers*.

Oddly Square

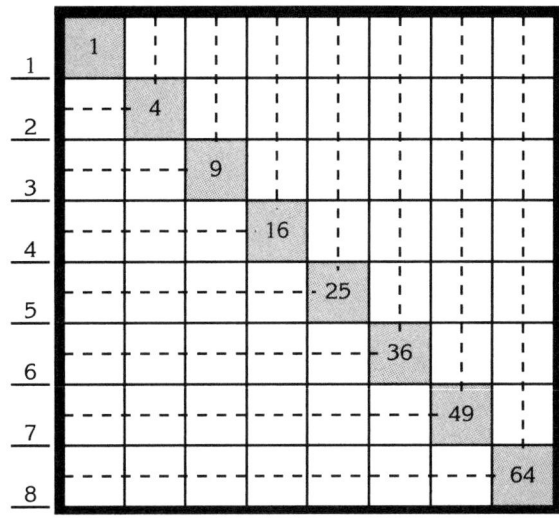

Oddly Square Patterns
- The sequence of numbers along the left side of the *Oddly Square* grid is {1, 2, 3, 4, 5, 6, 7, 8}.
- The sequence of numbers that appears along the diagonal is {1, 3, 5, 7, 9, 11, 13, 15}. These are known as the square numbers.
- The row number and the number of tiles in the row are the same.
- To compute the number of tiles in the square for any row, double the row number.

Table One		Table Two
Row	Row x Row	
1	1 x 1 = 1	1 = 1
2	2 x 2 = 4	1 + 3 = 4
3	3 x 3 = 9	1 + 3 + 5 = 9
4	4 x 4 = 16	1 + 3 + 5 + 7 = 16
5	5 x 5 = 25	1 + 3 + 5 + 7 + 9 = 25
6	6 x 6 = 36	1 + 3 + 5 + 7 + 9 + 11 = 36
7	7 x 7 = 49	1 + 3 + 5 + 7 + 9 + 11 + 13 = 49
8	8 x 8 = 64	1 + 3 + 5 + 7 + 9 + 11 + 13 + 15 = 64

* © Copyright 2010. National Governors Association Center for Best Practices and Council of Chief State School Officers. All rights reserved.

Fiddler Fun

Key Questions

1. What geometric patterns do the natural (counting) and odd numbers make?
2. What numerical series are formed from the natural and odd number sequences?

Learning Goal

explore the geometric patterns and numerical series formed by the natural (counting) and odd numbers.

Fiddler Fun
Making a Set of Fiddlers

Materials
- Area Tiles or pennies
- Transparent tape
- Half-inch self-adhesive dots
- Scissors
- Ruler

Procedure

1. Tape the ends of a ruler to a flat surface.

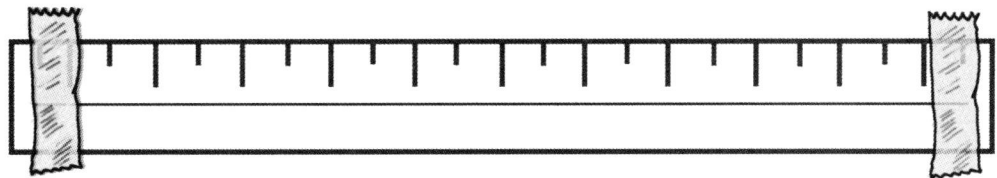

2. Align the tiles or pennies against the edge of the ruler.

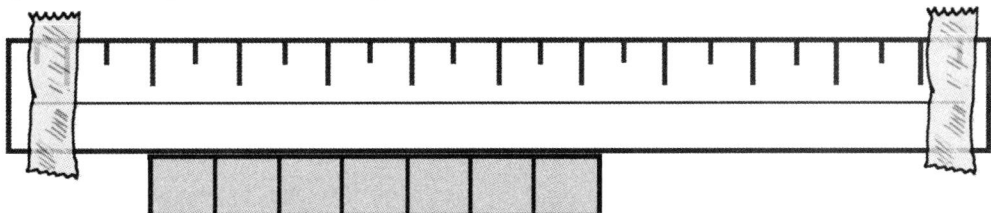

3. Cover the tiles with a single strip of transparent tape.

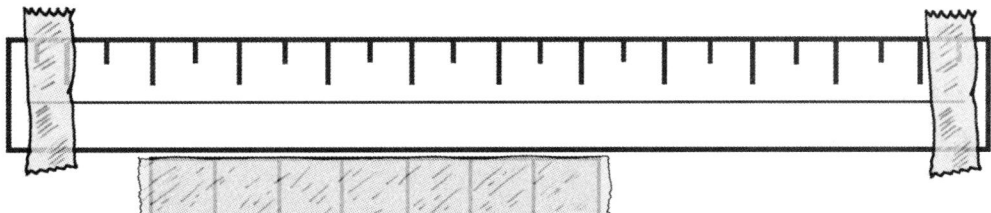

4. Trim away any excess tape with a pair of scissors.

5. Turn the tiles over and put a self-adhesive dot in the center of each one. Write the number of tiles in the Fiddler on one of the end dots.

6. Repeat this process to make a complete set of Fiddlers.

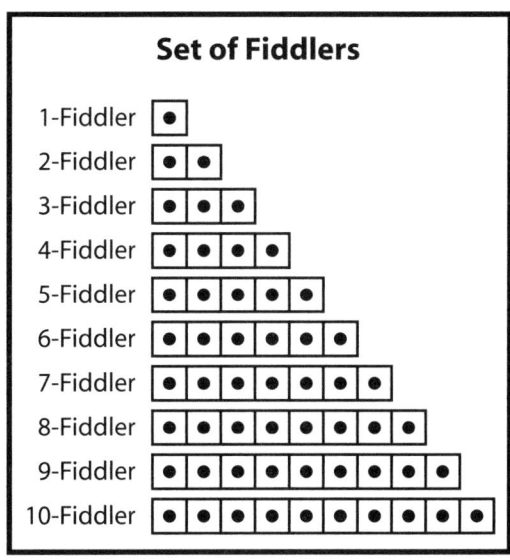

Fiddler Fun
Counting Fiddlers

1. Arrange your set of Fiddlers in a stair-step order.

2. Move the 10-Fiddler away from the group of Fiddlers.

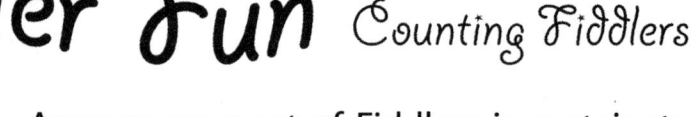
10–Fiddler

3. Using exactly two Fiddlers, form another group of 10. Draw a line connecting the two Fiddlers you used to form your group of 10.

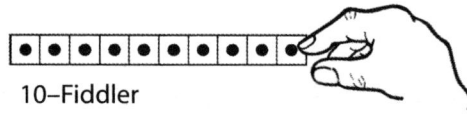
9–Fiddler

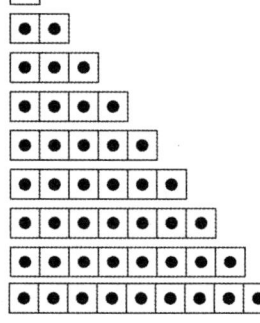

4. List the groups of 10 you can form using two Fiddlers.

(_____ + _____) = 10

(_____ + _____) = 10

(_____ + _____) = 10

(_____ + _____) = 10

(_____ + _____) = 10

5. What is the sum of the groups of 10?

6. What Fiddler wasn't used to form a group of 10?

7. What is the sum of all the Fiddlers?

SOLVE IT! 4th   © 2012 AIMS Education Foundation

Fiddler Fun — Fiddler Grid

Instructions

1. Player One: Put a 1-Fiddler in the *Start* square. Record "1" on the line to the left of the square.
2. Player Two: Put the next consecutive Fiddler (2-Fiddler) directly under the 1-Fiddler. Record "2" on the line to the left of the square.
3. Take turns and repeat this process through the bottom line of the grid.
4. Move the 1-Fiddler aside and record the sum (1) in the gray square.
5. Move the 2-Fiddler aside and record the sum of the 1- and 2-Fiddlers (3) in the second gray square.
6. Repeat this process until all gray squares are filled with sums.

SOLVE IT! 4th   29   © 2012 AIMS Education Foundation

Fiddler Fun — Number Patterns

1. Beginning with 1, write the sequence of numbers that appears along the left side of the grid.

   { _____ , _____ , _____ , _____ , _____ , _____ , _____ , _____ }

2. What is the rule that makes this number sequence?

3. What is the name of this number sequence?

4. Beginning with 1, write the series of numbers that appears along the diagonal.

   { _____ , _____ , _____ , _____ , _____ , _____ , _____ , _____ }

5. Remove all but the 1- and 2-Fiddlers from the grid. A triangle can be drawn that passes through every perimeter point.

   Add a 3-Fiddler to the grid. The pattern now looks like this. Draw a triangle on the pattern that passes through every perimeter point.

   Can you do the same for the patterns formed by adding a 4-Fiddler and then a 5-Fiddler?

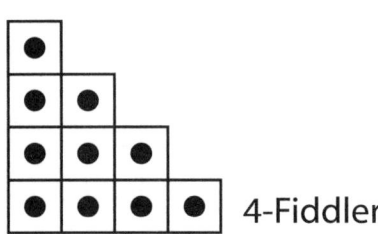 4-Fiddler   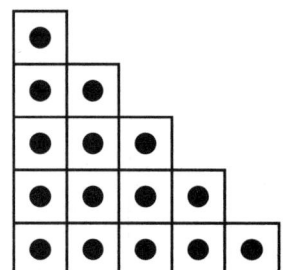 5-Fiddler

6. Why would triangular numbers be a good name for this number series?

SOLVE IT! 4th   30   © 2012 AIMS Education Foundation

Fiddler Fun *Oddly Square*

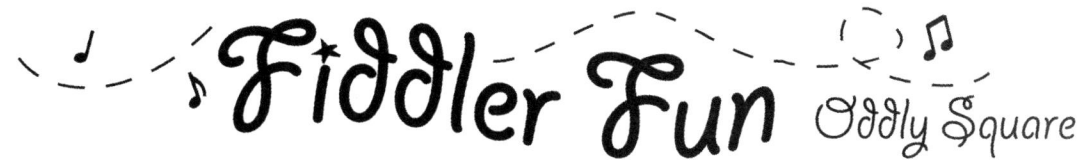

Instructions
1. First player places a 1-Fiddler in the square labeled *Start*.
2. Second player is challenged to make the next smallest possible square using two consecutive Fiddlers.
3. Taking turns, repeat this process through the bottom line of the grid.
4. Starting with the largest square, take turns removing the two consecutive Fiddlers used to make the square. Record the sum of the two Fiddlers in the shaded corner square.

SOLVE IT! 4th   31   © 2012 AIMS Education Foundation

Fiddler Fun — Oddly Square Patterns

1. Beginning with 1, write the sequence of numbers that appears along the left side of the grid.

   { _____ , _____ , _____ , _____ , _____ , _____ , _____ , _____ }

2. Beginning with 1, write the sequence of numbers that appears along the diagonal.

   { _____ , _____ , _____ , _____ , _____ , _____ , _____ , _____ }

3. What is the name of the sequence of numbers that appears along the diagonal?

4. Clear the *Oddly Square* grid of Fiddlers and, taking turns, rebuild the squares on the grid. Compare the row number with the number of tiles in the row. What do you notice?

5. If you know the row number, how would you compute the number of tiles in the square on that row?

6. Fill in *Table One* and *Table Two*.

Table One

Row	Row x Row
1	_____
2	_____
3	_____
4	_____
5	_____
6	_____
7	_____
8	_____

Table Two

1 = 1
1 + 3 = 4
1 + 3 + 5 = ___
1 + 3 + 5 + 7 = ___
1 + 3 + 5 + 7 + 9 = ___
1 + 3 + 5 + 7 + 9 + 11 = ___
1 + 3 + 5 + 7 + 9 + 11 + 13 = ___
1 + 3 + 5 + 7 + 9 + 11 + 13 + 15 = ___

7. Compare the results in *Tables One* and *Two*.

8. Why would square numbers be a good name for the number series formed by summing the odd numbers?

Connecting Learning

Part One—Introduction

1. What things did you discover about your Fiddlers?

2. What kinds of problems can you solve with your Fiddlers?

3. What do your Fiddlers tell you about the numbers they represent?

Part Two—Triangular Numbers

1. What are the triangular numbers?

2. How are they made?

3. Why do you think they are called triangular numbers?

Connecting Learning

4. What are the next two numbers in the series? Are these two numbers also triangular numbers? Explain.

5. Arrange all 10 Fiddlers on the grid page. How does the number of tiles down the left side, along the bottom, and along the diagonal compare?

6. If you had an 11-Fiddler to add to the bottom, would your answer to the comparison question change? Explain.

7. Using both sets of Fiddlers, what's the largest triangular number you can build?

8. If you were to continue adding larger Fiddlers in counting order, will each successive number in the series be a triangular number? Explain.

Connecting Learning

Part Three—Square Numbers

1. What are the square numbers?

2. How are they made?

3. Why do you think they are called square numbers?

4. Using both sets of Fiddlers, what's the largest square number you can build?

5. If the grid were as large as you wanted and you were to continue adding two consecutive Fiddlers in counting order, would each successive number in the series be a square number? Explain.

Problem-Solving Strategies
□ + □ = □ Write a Number Sentence

Sometimes it helps to write down the parts of a problem as a number sentence. Then you can see if what you are doing makes sense. Are you using the right numbers? Are you adding where you should be subtracting? If you have a number sentence, you can answer these questions.

Addition Rummy

Topic
Addition

Key Question
How can you use your cards to create correct addition problems?

Learning Goal
Students will play a game to reinforce addition and gain practice with integers.

Guiding Document
Common Core State Standards for Mathematics*
- *Make sense of problems and persevere in solving them. (MP1)*

Math
Number and operations
   addition
   integers
Problem solving

Integrated Processes
Observing
Comparing and contrasting
Recording

Problem-Solving Strategies
Write a number sentence
Use manipulatives

Materials
Addition Rummy cards (see *Management 3*)
Score cards

Background Information
   Brain research tells us that students' brains are enriched when they are put in situations combining challenge with interactive feedback. Games offer a logical and simple avenue to provide students with both of these elements. Games have an element of novelty; they are fun for students, and they provide immediate feedback—you win or you lose, you get points or you lose points, etc.
   This activity provides a game in which students will practice their addition facts and work with integers. The deck used for *Addition Rummy* consists of 54 cards containing three each of the numbers from zero to 11, six plus signs, and six equals signs. The play of the game is identical to that of Gin Rummy, but instead of laying down sets or runs of numbers, students will lay down correct addition problems.
   In addition to the value students will gain by simply practicing their addition facts in an engaging manner, they will also be working with integers as they keep score. The scorecard directs them to record both positive and negative points for each round, come up with a round total, and keep a running subtotal of the points acquired in multiple rounds. In every round, two out of three players will receive negative points, virtually guaranteeing that each student will have a chance to practice computation with integers.

Management
1. This activity is well-suited for a center or other small-group time where a few students can play it for short periods of time. It can also be done as a whole-class activity if desired.
2. A complete set of rules for *Addition Rummy* is included. You may choose to distribute the rules for students to read, or simply explain the rules to the class.
3. Copy the *Addition Rummy* playing cards onto cardstock and cut them out. You may also wish to laminate them for durability. The number of decks you need will be determined by how you choose to present the activity.
4. A page of scorecards is provided for you to copy and cut into fourths. Each student will need one scorecard for every game played.

Procedure
1. Distribute the materials to students and go over the rules for the game.
2. Provide time for students to play several rounds and record their scores.
3. Repeat as desired for additional reinforcement.

Connecting Learning
1. What strategies did you develop while playing this game?
2. Which cards did you most want to have? Why?
3. If you could change one of the rules for this game, what would it be? Why?

Extensions
1. Increase the values on the cards to make the computation more challenging.
2. Add subtraction signs to the deck.

* © Copyright 2010. National Governors Association Center for Best Practices and Council of Chief State School Officers. All rights reserved.

Addition Rummy

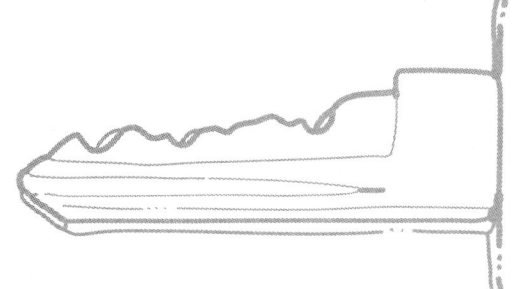

Key Question

How can you use your cards to create correct addition problems?

Learning Goal

Students will:

play a game to reinforce addition and gain practice with integers.

 **Three**

To arrange your cards to form true addition problems and be the first person to lay down all of your cards

1. Each player is dealt seven cards. The remaining cards are placed face down to form the draw pile. Flip over the top card to form the start of the discard pile. Player to the left of the dealer begins.
2. Each turn a player must draw one or more cards and discard one card. A player can draw from the draw pile or from the discard pile. More than one card may be drawn from the discard pile at a time; however, when picking up more than one card, the player must be able to immediately use the bottom card picked up.
3. When discarding, lay the cards next to each other so that the value of each discard can be seen.
4. When a player is able to create a true addition problem with the cards in his/her hand, he/she may lay those cards down, face up for the other players to see. Addition problems can take any form, as long as they are true. They may use some or all of the cards in the player's hand. For example: 4 + 5 = 6 + 3; 10 + 1 = 11; 2 + 3 + 4 = 9.
5. A player is not required to lay down an addition problem as soon as he/she creates one. It can sometimes be advantageous to hold on to cards until they can all be laid down at once. There is always the risk, however, that another player will go out first and all cards not laid down will be counted as negative points.
6. Once a player has laid down at least one equation he/she may play on the equations of the other players. In order to play on another addition problem, it must remain true without rearranging the numbers. For example, a player could add "+ 0" to either side of any addition problem without making it untrue.
7. The game ends when one player is able to get rid of all his/her cards. Note: A player must be able to discard to end the game. Scores are determined and recorded, and the player who went first becomes the new dealer. The first player to reach 100 points is the winner.

Players receive positive points for all cards they have laid down, and negative points for each card remaining in their hands. (The first player to go out receives no negative points.) Each number card is worth its face value. Plus and equals cards are each worth 15 points.

SOLVE IT! 4th © 2012 AIMS Education Foundation

Addition Rummy

Player: _____

Positive points	Negative points	Round score	Subtotal
			✕
		Total Score	

Addition Rummy

Player: _____

Positive points	Negative points	Round score	Subtotal
			✕
		Total Score	

Addition Rummy

Player: _____

Positive points	Negative points	Round score	Subtotal
			✕
		Total Score	

Addition Rummy

Player: _____

Positive points	Negative points	Round score	Subtotal
			✕
		Total Score	

SOLVE IT! 4th © 2012 AIMS Education Foundation

1	7	3
2	8	4
3	9	5
4	10	6
5	1	7
6	2	8

11	+	=
11	+	=
0	+	=
0	+	=
10	+	=
9	+	=

2	7	8
1	5	6
0	3	4
11	+	=
10	+	=
9	+	=

SOLVE IT! 4th   45   © 2012 AIMS Education Foundation

Addition Rummy

Connecting Learning

1. What strategies did you develop while playing this game?

2. Which cards did you most want to have? Why?

3. If you could change one of the rules for this game, what would it be? Why?

To A+D+D, or Not to Add?

Topic
Odd/even numbers

Key Question
What rules can you discover that will let you win the game every time?

Learning Goals
Students will:
- play a game in which they add and/or subtract a series of consecutive numbers,
- determine if the sums are odd or even, and
- discover what makes the sums odd or even.

Guiding Document
Common Core State Standards for Mathematics*
- *Make sense of problems and persevere in solving them. (MP1)*
- *Look for and make use of structure. (MP7)*

Math
Number and operations
    addition
    subtraction
    integers
    odd/even numbers
Problem solving

Integrated Processes
Observing
Collecting and recording data
Comparing and contrasting
Generalizing

Problem-Solving Strategies
Write a number sentence
Look for patterns

Materials
Student pages

Background Information
This activity is a variation of a game called *Plus or Minus* from an Internet math site. In *Plus or Minus*, you play against the computer, taking turns placing either a plus or a minus sign between a set of consecutive whole numbers starting with one. You can choose a sequence ranging between two and 20 numbers. When all of the spaces between numbers have been filled, the computer totals the expression. If the total is odd, you win; if the total is even, the computer wins.

In our version of the game, students play in pairs, with one person as player one, and one person as player two. On the first student page, they will take turns placing either a plus or a minus sign between pre-set number sequences as directed. If the total is odd, player one is the winner; if the total is even, player two is the winner. On the second student page, each player will get to make up a number sequence of his or her own, and each student will get a chance to go first on the same set of numbers.

As students have repeated experiences with various number sequences, they will begin to realize that it does not matter which player goes first, or even what symbol each person places between the numbers. The sole factor that determines whether the total will be odd or even is the number of odd numbers in the sequence. If there is an odd number of odd numbers, the total of the expression will always be odd. If there is an even number of odd numbers, the total will always be even. This can be illustrated by the following examples:

$2+3+4=9$ $2-3+4=3$ $2+3-4=1$ $2-3-4=-5$
$1+2+3=6$ $1-2+3=2$ $1+2-3=0$ $1-2-3=-4$

In the first number sequence (2, 3, 4), there is one odd number. As you can see, no matter what the combination of symbols is used, the total of the expression is always odd. In the second sequence (1, 2, 3), there are two odd numbers, which always results in an even total.

This explanation can be clarified by looking at odds and evens in general terms. When you add or subtract two odd numbers, the result is always even $(3 + 5 = 8, 9 - 5 = 4)$. When you add or subtract two even numbers, the result is always even $(2 + 4 = 6, 8 - 6 = 2)$. The only time you will have an odd sum or difference is when you add or subtract an odd number and an even number $(3 + 4 = 7, 9 - 6 = 3)$. It therefore follows that if you have a sequence of numbers to be added and/or subtracted, the only way the total will be odd is if there is an odd number of odd numbers.

SOLVE IT! 4th

Management

1. Because the sequences are made up of consecutive whole numbers and students may choose to add or subtract, a working understanding of integers is necessary for students to complete this activity as written.
2. Students will be guided to the discovery of this generalization by organizing their results in a table and answering questions. If your class does well with open-ended problems, you can challenge them to develop the generalization without giving them the accompanying pages.

Procedure

1. Have students get into pairs and distribute one copy of the first student page to each pair.
2. Go over the rules and allow time for students to complete the page.
3. Distribute the second student page to each group and allow time for them to come up with their own sequences and play.
4. When groups are done, give each student a copy of the third and fourth student pages. Allow them to work together to complete the table and answer the questions.
5. Conduct a time of class discussion where students share the generalizations they discovered.
6. If desired, assess their understanding by playing a modified version of the game as a class. Write several different sequences of numbers on the chalkboard or the overhead projector and have students determine who will win each game without actually playing them. Assume that you are player one (odd total wins) and they are player two (even total wins).

Connecting Learning

1. Which sequences gave you an odd total? Why?
2. Which sequences gave you an even total? Why?
3. Does the number of plus or minus signs affect whether the total will be odd or even? [no] Why or why not? (See *Background Information*.)
4. Does the number of odd numbers or even numbers affect whether the total will be odd or even? [yes] Why or why not? (See *Background Information*.)
5. What do the sequences that give you odd totals have in common? [They all have an odd number of odd numbers.]
6. What do the sequences that give you even totals have in common? [They all have an even number of odd numbers.]
7. What generalization did you develop for determining who will win the game?

Extensions

1. Use the extension page provided to examine the patterns relating to the total based on how many numbers are in the sequence. Students will discover that when there is an odd number of numbers in the sequence, it is possible for either player to win, depending on the numbers. However, when there is an even number of numbers in the sequence, one player will win every time.
2. Use negative numbers in the sequences.
3. Eliminate the restriction that the numbers must be consecutive.

Internet Connections

Cut the Knot
http://www.cut-the-knot.org
For the online version of this puzzle follow the "Games and Puzzles" link and click on *Plus or Minus*.

* © Copyright 2010. National Governors Association Center for Best Practices and Council of Chief State School Officers. All rights reserved.

To A+D+D, or Not to Add?

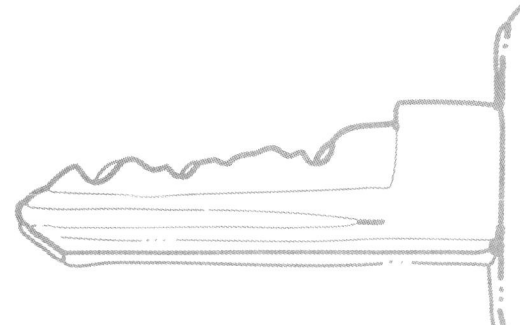

Key Question

What rules can you discover that will let you win the game every time?

Learning Goals

Students will:

- play a game in which they add and/or subtract a series of consecutive numbers,

- determine if the sums are odd or even, and

- discover what makes the sums odd or even.

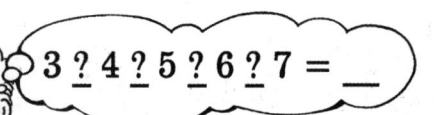

To A+D+D, or Not to Add?

Choose one person to be player one, and one person to be player two. Player one begins by placing either a plus or a minus sign between two of the numbers. Player two then places either a plus or a minus sign between the remaining two numbers. Find the total for the expression. If the total is odd, player one is the winner. If the total is even, player two is the winner.

1. Player one begins.

         1   2   3 =

   Was the sum odd or even?

   Winner:

2. Player two begins.

         2   3   4 =

   Was the sum odd or even?

   Winner:

3. Player one begins.

         1   2   3   4 =

   Was the sum odd or even?

   Winner:

4. Player two begins.

         2   3   4   5 =

   Was the sum odd or even?

   Winner:

To A+D+D, or Not to Add?

Use the same rules as you did for the problems on the previous page.

1. Player one writes a sequence of numbers below. The sequence should be different from those on the first page. You may use as many numbers as you want, but they must be consecutive whole numbers. Have player one begin.

   Was the sum odd or even?

   Winner:

2. Use the same sequence of numbers. Have player two begin.

   Was the sum odd or even?

   Winner:

3. Player two writes a different sequence of numbers. Have player one begin.

   Was the sum odd or even?

   Winner:

4. Use the same sequence of numbers. Have player two begin.

   Was the sum odd or even?

   Winner:

To A+D+D, or Not to Add?

Fill in the table below to help you analyze your results.

Numbers	# of + signs	# of − signs	# of odd numbers	# of even numbers	Total: odd/even

1. Does the number of plus or minus signs affect whether the total will be odd or even? Why or why not?

2. Does the number of odd numbers or even numbers affect whether the total will be odd or even? Why or why not?

3. Which player won more of the games you played? Why?

To A+D+D, or Not to Add?

1. Which sequences gave you an odd total?

2. Try these same sequences using at least two different combinations of symbols. What results do you get?

3. Which sequences gave you an even total?

4. Try these same sequences using at least two different combinations of symbols. What results do you get?

5. What do the sequences that give you odd totals have in common?

6. What do the sequences that give you even totals have in common?

7. Based on what you have discovered, make a generalization that will let you know which player will win before the game begins.

8. Use this information to write a sequence that would let you win every time if you were player one. Write a second sequence that would let you win every time if you were player two.

To A+D+D, or Not to Add?

Extension

This table will help you analyze the effect of the amount of numbers in the sequence on the total. Use what you have learned to complete the missing information.

Total numbers in sequence	Odd and/or even totals?	Example(s)
2	Odd	1 + 2 = 3, 5 − 6 = -1
3	Odd and Even	
4		
5		
6		
7		
8		
9		

1. Describe the patterns you see in the table.

2. Make a generalization that explains these patterns.

3. How would the results change if the numbers in the sequences did not have to be consecutive?

To A+D+D, or Not to Add?

Connecting Learning

1. Which sequences gave you an odd total? Why?

2. Which sequences gave you an even total? Why?

3. Does the number of plus or minus signs affect whether the total will be odd or even? Why or why not?

4. Does the number of odd numbers or even numbers affect whether the total will be odd or even? Why or why not?

5. What do the sequences that give you odd totals have in common?

6. What do the sequences that give you even totals have in common?

7. What generalization did you develop for determining who will win the game?

Topic
Number combinations

Key Question
How many ways can you mail a letter for ¢37 using the 10 national parks stamps available in 1934?

Learning Goals
Students will:
- use the numbers from one to 10 to make combinations that sum 37, and
- try to determine if they have found all possible solutions.

Guiding Document
*Common Core State Standards for Mathematics**
- *Make sense of problems and persevere in solving them. (MP1)*
- *Look for and make use of structure. (MP7)*

Math
Number and operations
    addition
Problem solving

Integrated Processes
Observing
Comparing and contrasting
Recording
Analyzing

Problem-Solving Strategies
Write a number sentence
Look for patterns
Organize the information
Use manipulatives

Materials
Set of stamps, one per group
Chart paper
Student pages

Background Information
In 1934, the United States Postal Service issued 10 stamps commemorating the national parks. The following parks were represented: Yosemite (El Capitan, 1¢), Grand Canyon (2¢), Mount Rainier (Mirror Lake, 3¢), Mesa Verde (Cliff Palace, 4¢), Yellowstone (Old Faithful, 5¢), Crater Lake (6¢), Acadia (Great Head, 7¢), Zion (Great White Throne, 8¢), Glacier (Mount Rockwell and Two Medicine Lake, 9¢), and Great Smoky Mountains (10¢). We have created our own replications based on these original stamps that students will use to solve a modern-day problem: How many different ways can you mail a letter for 37¢ (the cost 70 years later in 2004) using one set of the 1934 national parks stamps? This problem may seem trivial at first glance, but a closer look reveals that multiple solutions, patterns, and systematic problem solving are involved.

Management
1. Students should be allowed to work on this activity in groups, and will likely need several days to complete it.
2. Each group should be given one set of national parks stamps to help them as they search for solutions.

Procedure
1. Have students get into groups and distribute the stamps and first student page. Go over the problem and be sure students understand the challenge and the restrictions.
2. Ask the *Key Question* and have students guess how many solutions they think can be found. Record these guesses on the top of a sheet of chart paper to be returned to later.
3. Allow time for students to work on the problem in their groups. What they should begin to realize is that they will need to develop some sort of systematic method for organizing their solutions in order to determine if they have found every possibility.
4. When students believe they have found all of the possible solutions, give them the second student page and follow up with a time of class discussion.

5. Have each group share their solutions, and record them on the sheet of chart paper to make a class "master" list. If fewer than 27 solutions were discovered, search together to find those that are still missing. Compare the actual number of solutions discovered to students' initial guesses.

Connecting Learning
1. How many solutions did you guess were possible? How many are actually possible? [27]
2. What is the fewest number of stamps that you can use to mail a letter? [five] ...the greatest number? [eight]
3. Which stamps did you use most often in your solutions? Why do you think this is?
4. Which stamps did you use least often in your solutions? Why do you think this is?
5. How did your group choose to organize its solutions? Why did you choose this method?
6. How does your group's method compare to those of other groups?

Extensions
1. Change the total to the actual cost of currently mailing a letter.
2. Change the values of some of the stamps and compare the solutions. For example, change the value of the Yosemite stamp from 1¢ to 11¢.
3. Pick two or three stamps and see how many ways they could be used to create 37¢ if each one could be used more than once.
4. Determine stamp values for a three- or four-stamp combination that only has one possible way to make 37¢.

Solutions
There are 27 unique solutions to this problem that we are aware of. Of those, three are five-stamp combinations, 13 are six-stamp combinations, 10 are seven-stamp combinations, and one is an eight-stamp combination.

Stamp combinations that total 37¢:
10¢ + 9¢ + 8¢ + 7¢ + 3¢
10¢ + 9¢ + 8¢ + 7¢ + 2¢ + 1¢
10¢ + 9¢ + 8¢ + 6¢ + 4¢
10¢ + 9¢ + 8¢ + 6¢ + 3¢ + 1¢
10¢ + 9¢ + 8¢ + 5¢ + 4¢ + 1¢
10¢ + 9¢ + 8¢ + 5¢ + 3¢ + 2¢
10¢ + 9¢ + 8¢ + 4¢ + 3¢ + 2¢ + 1¢
10¢ + 9¢ + 7¢ + 6¢ + 5¢
10¢ + 9¢ + 7¢ + 6¢ + 4¢ + 1¢
10¢ + 9¢ + 7¢ + 6¢ + 3¢ + 2¢
10¢ + 9¢ + 7¢ + 5¢ + 4¢ + 2¢
10¢ + 9¢ + 7¢ + 5¢ + 3¢ + 2¢ + 1¢
10¢ + 9¢ + 6¢ + 5¢ + 4¢ + 3¢
10¢ + 9¢ + 6¢ + 5¢ + 4¢ + 2¢ + 1¢
10¢ + 8¢ + 7¢ + 6¢ + 5¢ + 1¢
10¢ + 8¢ + 7¢ + 6¢ + 4¢ + 2¢
10¢ + 8¢ + 7¢ + 6¢ + 3¢ + 2¢ + 1¢
10¢ + 8¢ + 7¢ + 5¢ + 4¢ + 3¢
10¢ + 8¢ + 7¢ + 5¢ + 4¢ + 2¢ + 1¢
10¢ + 8¢ + 6¢ + 5¢ + 4¢ + 3¢ + 1¢
10¢ + 7¢ + 6¢ + 5¢ + 4¢ + 3¢ + 2¢
9¢ + 8¢ + 7¢ + 6¢ + 5¢ + 2¢
9¢ + 8¢ + 7¢ + 6¢ + 4¢ + 3¢
9¢ + 8¢ + 7¢ + 6¢ + 4¢ + 2¢ + 1¢
9¢ + 8¢ + 7¢ + 5¢ + 4¢ + 3¢ + 1¢
9¢ + 8¢ + 6¢ + 5¢ + 4¢ + 3¢ + 2¢
9¢ + 7¢ + 6¢ + 5¢ + 4¢ + 3¢ + 2¢ + 1¢

* © Copyright 2010. National Governors Association Center for Best Practices and Council of Chief State School Officers. All rights reserved.

PARKS POSTAGE PROBLEM

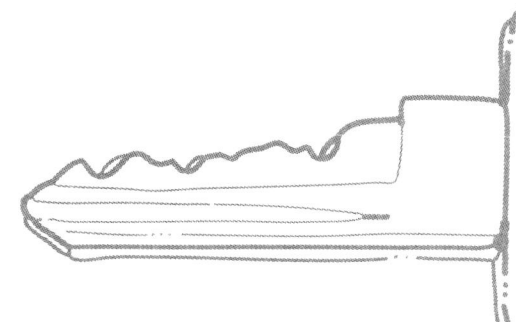

Key Question

How many ways can you mail a letter for ¢37 using the 10 national parks stamps available in 1934?

Learning Goals

Students will:

- use the numbers from one to 10 to make combinations that sum 37, and

- try to determine if they have found all possible solutions.

National Parks Stamps
Cut out one set of stamps for each group of students.

SOLVE IT! 4th © 2012 AIMS Education Foundation

PARKS POSTAGE PROBLEM

In 1934, the U.S. Postal Service issued 10 stamps to commemorate the national parks. The stamps had values from 1¢ to 10¢. Each had a picture of a different national park. At the time, it cost 3¢ to mail a first class letter in the United States. In 2004—70 years later—it cost 37¢.

Challenge
How many different ways can you use the 10 national parks stamps from 1934 to mail a letter at 2004 rates?

You may only use each stamp once per solution. For example, you could use 10¢ + 9¢ + 7¢ + 6¢ + 4¢ + 1¢, but you could not use 10¢ + 10¢ + 10¢ + 7¢. Use the front and/or back of this page to record your solutions.

SOLVE IT! 4th 61 © 2012 AIMS Education Foundation

PARKS POSTAGE PROBLEM

Answer these questions once your group has found all of the solutions it can.

1. How many solutions did your group discover? Do you think you have found them all? Why or why not?

2. What method(s) did your group use to organize your solutions? How did this help?

3. What is the fewest number of stamps you can use to make 37¢? What is the greatest number of stamps you can use?

4. Which stamp appeared most often in your solutions? Why?

5. Which stamp appeared least often in your solutions? Why?

6. What other questions do you have now that you have done this activity?

Connecting Learning

1. How many solutions did you guess were possible? How many are actually possible?

2. What is the fewest number of stamps that you can use to mail a letter? …the greatest number?

3. Which stamps did you use most often in your solutions? Why do you think this is?

4. Which stamps did you use least often in your solutions? Why do you think this is?

5. How did your group choose to organize its solutions? Why did you choose this method?

6. How does your group's method compare to those of other groups?

Problem-Solving Strategies
Draw out the Problem

Drawing pictures is a useful problem-solving tool. Pictures help you keep track of important information. They also help when a problem has lots of details. With a picture, you can see every part of the problem at once. Keep the pictures simple. You don't need to spend lots of time drawing.

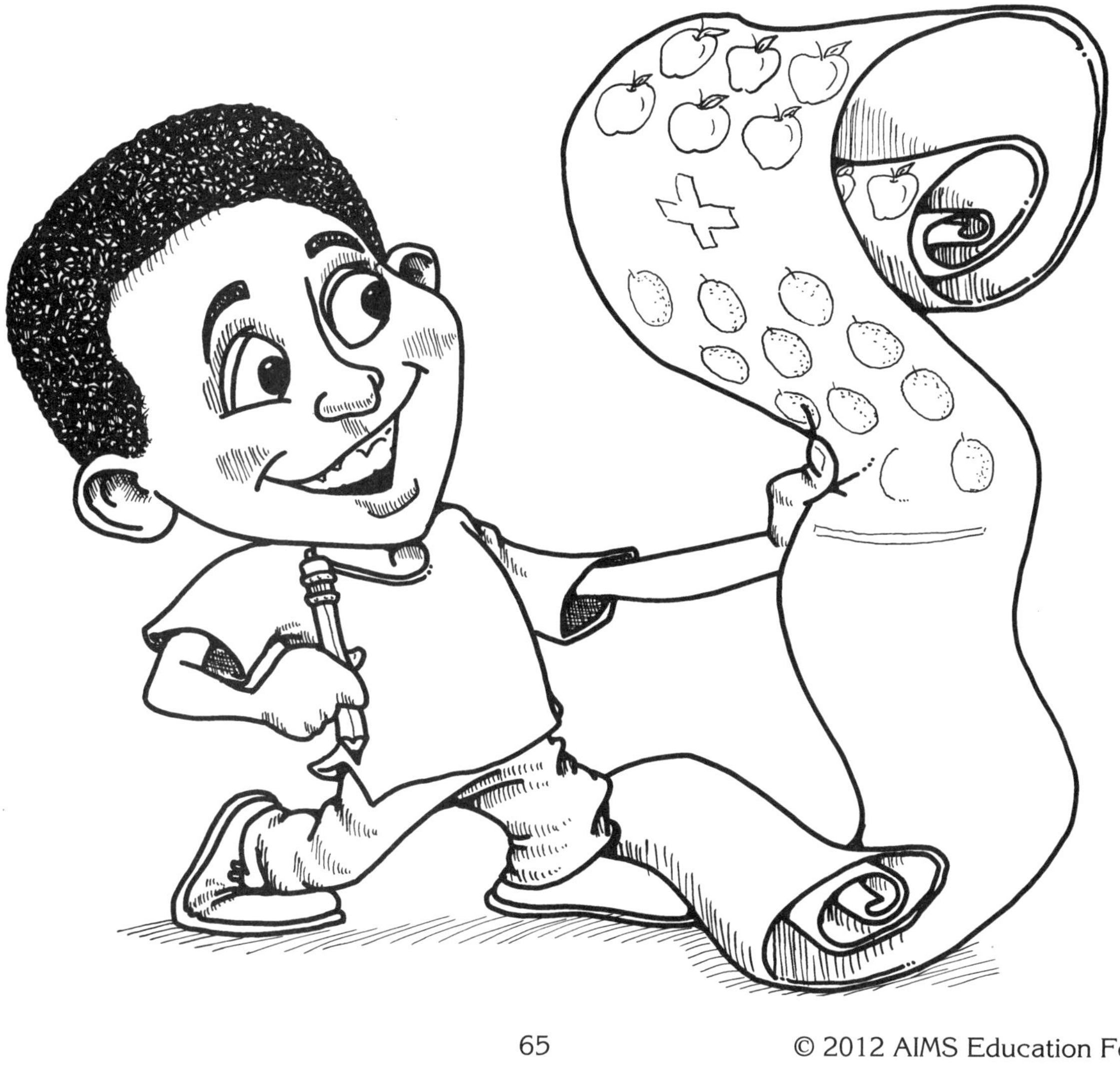

Key Counts

Topic
Problem solving

Key Question
What is the maximum number of tries it could take to match five unmarked keys to five cars?

Learning Goals
Students will:
- determine the maximum number of tries it could take to match five keys to five cars, and
- use what they learned while doing this to determine the maximum number of tries it could take to match 10 keys and 10 cars.

Guiding Document
- *Make sense of problems and persevere in solving them. (MP1)*
- *Construct viable arguments and critique the reasoning of others. (MP3)*

Math
Problem solving

Integrated Processes
Observing
Comparing and contrasting
Recording
Generalizing

Problem-Solving Strategies
Draw out the problem
Look for patterns
Organize the information

Materials
Student page

Background Information
This activity is an open-ended mathematical investigation. The real value in this type of activity is the mathematical thinking and communicating it fosters in students as they work on the problem(s) at hand. Since students may use a variety of different approaches and problem-solving strategies while working to find their solutions, it is critical that a time for them to share their approaches and strategies in a whole-class session is built into the lesson.

In this problem, students are presented with the following scenario: using only trial and error, five unmarked keys must be matched with five cars. While very unlikely (one chance out of 120), this matching can be done in five tries—the minimum number. The maximum number of tries, which is equally unlikely, is 15—five tries to match the key for the first car, four tries for the second car, three for the third, two for the fourth, and one for the fifth.

As students discover the pattern used to find the maximum number of tries, they will be challenged to apply this knowledge to determining the maximum number of tries for 10 keys and 10 cars.

Management
1. Students should be allowed to work together in groups.
2. Encourage the use of a variety of problem-solving strategies. Make manipulatives available for students who may want to use them.

Procedure
1. Distribute the student page and have students get into groups.
2. Go over the instructions and allow time for groups to come up with their answers.
3. Close with a time of class discussion where groups share their answers, the methods they used to arrive at those answers, and their guesses for 10 keys and 10 cars.

Connecting Learning
1. What is the maximum number of tries it could take for the dealer to match the keys to the cars? [15]
2. How did you come up with this number?
3. How does the method your group used compare to the methods other groups used?
4. Do you think one method is more effective than others? Why or why not?
5. What is the maximum number of tries it would take to match 10 keys with 10 cars? [55] How do you know?
6. How would you find the maximum number of tries for 15 keys and cars? ...50 keys and cars?

Extensions
1. Increase the number of keys without increasing the number of cars. For example, if you have seven keys and five cars, what is the maximum number of tries it could take to match the keys? How does this number compare to the number for five keys? ...for six keys?
2. Write the generalization for the maximum number of tries in algebraic form.

* © Copyright 2010. National Governors Association Center for Best Practices and Council of Chief State School Officers. All rights reserved.

Key Question

What is the maximum number of tries it could take to match five unmarked keys to five cars?

Learning Goals

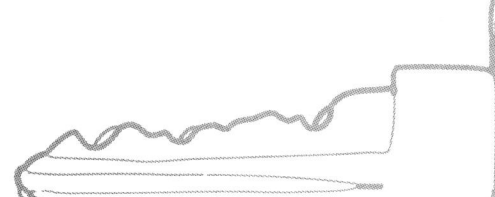

- determine the maximum number of tries it could take to match five keys to five cars, and

- use what they learned while doing this to determine the maximum number of tries it could take to match 10 keys and 10 cars.

Key Counts

The local car dealer just received five new cars on his lot. Unfortunately, the keys to the cars are unlabeled with no way to tell which key matches which car. If he got really lucky, he could match the keys to the cars in just five tries. (Do you see how this could be done?) If, however, he was unlucky, it would take more than five tries. What is the maximum number of tries he could make to match each key to its car?

Show your work in the space below. Be ready to share your problem-solving strategies with others in the class.

Using what you learned from the above problem, find the maximum number of tries it would take to match 10 keys with 10 cars. Show your work on the back of this paper.

Connecting Learning

1. What is the maximum number of tries it could take for the dealer to match the keys to the cars?

2. How did you come up with this number?

3. How does the method your group used compare to the methods other groups used?

4. Do you think one method is more effective than others? Why or why not?

5. What is the maximum number of tries it would take to match 10 keys with 10 cars? How do you know?

6. How would you find the maximum number of tries for 15 keys and cars? …50 keys and cars?

Finding Floors and Reckoning Rungs

Topic
Problem solving

Key Question
How can you solve the story problems presented on these pages?

Learning Goal
Students will use the problem-solving strategy of drawing a picture or diagram to solve the problems posed.

Guiding Document
*Common Core State Standards for Mathematics**
- *Make sense of problems and persevere in solving them. (MP1)*

Math
Problem solving

Integrated Processes
Observing
Comparing and contrasting
Relating
Generalizing

Problem-Solving Strategy
Draw out the problem

Materials
Student pages
Overhead transparency, optional

Background Information
In this activity, students are challenged with a number of similar story problems. While these problems can be solved in several different ways, drawing a picture or diagram resembling a vertical number line best solves them.

Management
1. Overhead transparencies of the student pages can be used to introduce the activity.
2. If students have not done story problems like the ones presented, it will help to do a problem or two together as a class.
3. Students can work on this activity individually or as part of a group.
4. The last problem on the second page is much more difficult than the previous problems. This problem can be done as a whole-class, problem-solving endeavor.

Procedure
1. Distribute the student pages and make sure students understand the problems.
2. Have students work on the problems.
3. Monitor students' work, facilitating the problem-solving process as necessary.
4. Have students use the backs of the papers to make up their own problems to solve.
5. When students have completed the problems, have a whole-class discussion where students share their solutions and solution processes.

Connecting Learning
Finding Floors
1. How many floors did Felicity go down? [3 floors]
2. How many floors are in Fernando's hotel? [9 floors]
3. How many floors do Faye and her friends have to go down to exit the building on the first floor? [8 floors]

Reckoning Rungs
1. How many rungs are on Leon's ladder? [6 rungs]
2. How many rungs are on Leticia's ladder? [8 rungs]
3. How many rungs are on Larry's ladder? [11 rungs]
4. How did you solve these problems?

* © Copyright 2010. National Governors Association Center for Best Practices and Council of Chief State School Officers. All rights reserved.

Finding Floors and Reckoning Rungs

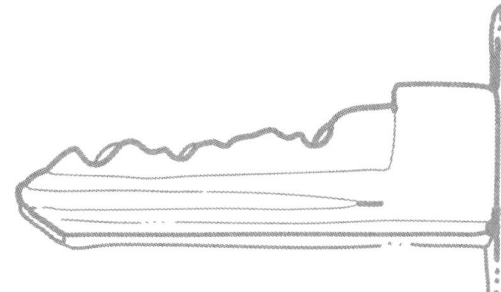

Key Question

How can you solve the story problems presented on these pages?

Learning Goal

Students will:

use the problem-solving strategy of drawing a picture or diagram to solve the problems posed.

Finding Floors

Try to solve each of the following elevator problems. You can draw a picture or diagram to help you find the solutions.

1. Felicity is on the first floor of a department store. She gets on the elevator and goes to the toy section on the third floor. Then she goes down one floor to the girls' clothing section. Next, she gets on the elevator and goes up four floors to school supplies. If she wants to get back to the toy floor, how many floors will she go down?

2. Fernando is staying at a hotel. He goes to his room on the fifth floor. He then goes down three floors to check out the swimming pool. Next, he goes up seven floors to the restaurant on the top floor. How many floors are there in the hotel?

3. Faye and her friends are on the fourth floor of the Federal Building. Together they go up two floors on the elevator, and then down four. They then go up seven floors. How many floors do they have to go down to exit the building on the first floor?

Challenge: Make up your own elevator problems on the back. After solving your problem, give it to a friend to try.

SOLVE IT! 4th © 2012 AIMS Education Foundation

Reckoning Rungs

Try to solve each of the following ladder problems. You can draw a picture or diagram to help you find the solutions.

1. Leon is on the third rung of a ladder. He goes down one rung. He then climbs four rungs to the top. How many rungs are on the ladder?

2. Leticia is on the fifth rung of a ladder. She climbs up two rungs. Then she goes down four rungs before climbing up five rungs to the top. How many rungs are there on the ladder?

3. Larry is on the middle rung of a ladder. He climbs up three rungs before climbing down six. Then he climbs up eight rungs to the top. How many rungs are on this ladder? (Be careful, this one is tricky.)

Challenge: Make up your own ladder problems on the back. After solving your problem, give it to a friend to try.

Connecting Learning

Finding Floors

1. How many floors did Felicity go down?

2. How many floors are in Fernando's hotel?

3. How many floors do Faye and her friends have to go down to exit the building on the first floor?

Finding Floors and Reckoning Rungs

Connecting Learning

Reckoning Rungs

1. How many rungs are on Leon's ladder?

2. How many rungs are on Leticia's ladder?

3. How many rungs are on Larry's ladder?

4. How did you solve these problems?

Polar Passage

Topic
Problem solving

Key Question
How can you make a six-day trek across Alaska when you can only carry four days' worth of supplies?

Learning Goal
Students will determine the minimum number of friends needed to help them complete a six-day trek across Alaska when each person can only carry four days' worth of supplies.

Guiding Document
Common Core State Standards for Mathematics*
- *Make sense of problems and persevere in solving them. (MP1)*
- *Construct viable arguments and critique the reasoning of others. (MP3)*
- *Use appropriate tools strategically. (MP5)*

Math
Logical thinking
Problem solving

Integrated Processes
Observing
Organizing
Recording

Problem-Solving Strategies
Draw out the problem
Use logical thinking
Use manipulatives

Materials
Student page

Background Information
Polar Passage is a logic problem that may seem familiar to some because of its similarities to *Desert Crossings*, which can be found in *Problem Solving: Just for the Fun of It! Book One*. Both activities challenge students to determine how to cross a given distance when they can only carry enough food/supplies to make it part of the way. However, upon closer examination, it can be seen that the similarities are not quite as great as they first seem; the solutions to the two activities are actually quite different, although based on the same general principle. In *Desert Crossings* students must leave food along the way and come back for it, while in *Polar Passage* students must use friends to help carry supplies.

In *Polar Passage*, students are challenged to determine the best way to make a six-day trek across Alaska when they can only carry four days' worth of supplies. They are told that they must ask friends to help them, but they must determine the minimum number of friends needed and how to carry out the trek.

As with *Desert Crossings*, this problem may yield many creative, but incorrect solutions such as "rent a snowmobile" or "use a dogsled." While these solutions give you insight into your students' creative processes, they do not help students develop the logical thinking skills that are critical to solve the problem. The key leap of insight necessary is for students to realize that the friends cannot go on the entire trek—they will have to go part of the way and return to Anchorage, leaving some of their supplies with the remaining traveler(s). Once this leap of insight is made, students can begin to experiment with different possibilities to get the fewest number of friends needed.

Management
1. Students should work on this problem in small groups, discussing ideas and developing plans.
2. Be sure to solve this problem yourself before giving it to your students. This will allow you to be a better facilitator and ask questions that guide discovery without giving away the answer.

Procedure
1. Distribute the student page and have students get into groups.
2. Allow time for students to work together to develop plans and solutions. Your students may come up with many solutions that work, but are not the best possible solution. Be sure to encourage all of these attempts and validate their answers.
3. Conduct a time of class discussion where groups share their responses. If none of the groups have come up with the best possible solution, you may wish to have a time of whole-class discovery where all of the students work together to determine the minimum number of friends needed.

SOLVE IT! 4th   77   © 2012 AIMS Education Foundation

Connecting Learning
1. How is it possible to make a six-day trek when you can only carry four days' worth of supplies? [have friends help you]
2. What is the minimum number of friends needed? [two]
3. Describe how you can use two friends to make the trek. (See *Solutions*.)
4. How did your group solve the problem? How does this compare to the methods used by other groups?

Solutions
The minimum number of friends needed to make the trek is two. It can be done as follows: You set out with two friends from Anchorage; each of you is carrying four days' worth of supplies. At the end of the first day, you each have three days' worth left. On the second day, one friend returns to Anchorage, using one days' worth of supplies; the other two days' worth are left with you and your remaining companion—giving you both four days' worth again. At the end of the second day you and your friend both have three days' worth of supplies left. On the third day, your remaining friend returns to Anchorage using two days' worth of supplies, and leaving the third days' worth with you—once again giving you a total of four days' worth. From there, you have only four days left to travel before reaching your grandmother, where you will have lots of food and holiday cookies waiting.

* © Copyright 2010. National Governors Association Center for Best Practices and Council of Chief State School Officers. All rights reserved.

POLAR PASSAGE

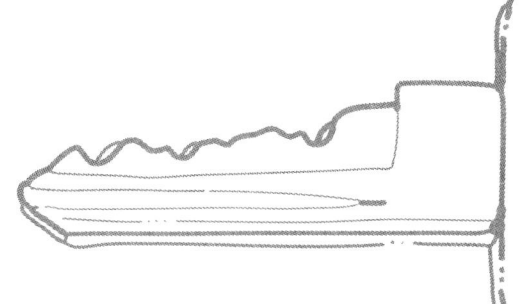

Key Question

How can you make a six-day trek across Alaska when you can only carry four days' worth of supplies?

Learning Goal

Students will:

determine the minimum number of friends needed to help them complete a six-day trek across Alaska when each person can only carry four days' worth of supplies.

POLAR PASSAGE

You have decided to visit your grandmother for the holidays. She lives in a remote village in Alaska. During the winter months they do not plow the roads, so the only way to get to the village is to snowshoe in. From your starting place in Anchorage it is a six-day trek. You must carry all of your food and supplies with you; however, you can only carry four days' worth of supplies in your pack. You will need to ask some friends to go with you to help you carry supplies. (Each friend can also carry only four days' worth of supplies.) What is the minimum number of friends that you need to help you make the trek? How will you do it?

Use the space above to describe your strategy in words and/or pictures.

POLAR PASSAGE

Connecting Learning

1. How is it possible to make a six-day trek when you can only carry four days' worth of supplies?

2. What is the minimum number of friends needed?

3. Describe how you can use two friends to make the trek.

4. How did your group solve the problem? How does this compare to the methods used by other groups?

Problem-Solving Strategies — Work Backwards

Sometimes it's best to start at the finish when solving a problem. Working backwards helps when you know the answer, but don't know how to get there. You can start at the end and find the missing steps to get to the beginning.

Topic
Probability

Key Question
Which color bear are you most likely to draw from a bag?

Learning Goals
Students will:
- predict the color of the bear they are likely to draw from a bag,
- draw and record the color of bear they actually draw, and
- evaluate the likelihood of certain events occurring.

Guiding Document
Common Core State Standards for Mathematics*
- *Make sense of problems and persevere in solving them. (MP1)*
- *Model with mathematics. (MP4)*

Math
Data analysis
  probability
Problem solving

Integrated Processes
Observing
Predicting
Collecting and recording data
Interpreting

Problem-Solving Strategies
Work backwards
Guess and check

Materials
Teddy Bear Counters, green and yellow
  (see *Management 2*)
Paper lunch sacks
Colored pencils, green and yellow
Student pages

Background Information
This activity provides a way for students to explore basic probability concepts and evaluate the likelihood of simple events occurring. They are given a bag of bears and told the number of each color that it contains. With this knowledge, they can predict, with some degree of accuracy, which color bear they are likely to draw. Because bears are not replaced once they are removed from the bag, the probability of drawing one color over another changes with each person's turn. As students predict and record each turn, they will begin to see how probability works in many real-world situations. Even though it may be very likely that something will happen, if it is not certain, it cannot be relied upon.

Management
1. Prepare enough different bags of bears so that each pair or group of students can have one. (Groups should have a maximum of four players; fewer are preferable.) Be sure to use bags that students cannot see through. Paper lunch sacks work well. Each bag should have a total of 12 bears to correspond to the number of spaces on the student pages. In some bags include more green, in some include more yellow, and in some include an equal number of green and yellow. Be sure there is at least one bear of each color in each bag. Suggested bag compositions are given here.

Bag One	Bag Two	Bag Three
1 yellow 11 green	3 yellow 9 green	4 yellow 8 green
Bag Four	**Bag Five**	**Bag Six**
5 yellow 7 green	6 yellow 6 green	7 yellow 5 green
Bag Seven	**Bag Eight**	**Bag Nine**
8 yellow 4 green	9 yellow 3 green	11 yellow 1 green

2. If you do not have Teddy Bear Counters, you may use other objects instead, as long as they are uniform and of two different colors—green and yellow.
3. Students will need one copy of the first student page for each different bag they use. Groups should use at least two or three different bags.

SOLVE IT! 4th © 2012 AIMS Education Foundation

Procedure
1. Introduce the activity by telling students the following story:
   The *Chewy Candies Café* sells bags of 12 gummy bears for $1.50. They only make two colors of gummy bears—green and yellow. Every Friday they have a special deal. They will fill your bag with 12 gummy bears—some yellow and some green. They will even tell you how many bears are green and how many are yellow. If you can correctly guess which color bear you will pull out of the bag first, you get the bag for free.
2. Tell students that they are going to be customers at the *Chewy Candies Café* who are trying to get their gummy bears for free. Instead of just trying to guess the color of the first bear drawn from the bag, they will guess the color of every bear they draw from the bag.
3. Distribute the first student page to each student and yellow and green colored pencils and one bag of bears to each pair or group of students.
4. Inform groups of the contents of their bags and have them record the numbers of yellow and green bears by writing down the numbers and coloring in the bears.
5. Describe the process by which students are to record their predictions and actual colors drawn. Before drawing from the bag, each player must guess whether he or she will draw a green bear or a yellow bear. This guess should be based on the knowledge of the contents of the bag and what has already been drawn. For example, if students are playing with a bag of three yellow bears and nine green bears, it is reasonable to guess that on the first turn a green bear will be drawn from the bag. If, after several turns, no yellow bears have been drawn, a student would be able to further conclude that drawing a yellow bear is now more likely because the number of green and yellow bears is closer to equal.
6. Explain that each player is to make a record of his or her guess by coloring in the *Guess* bears either yellow or green. Once a player has made and recorded a guess, he or she is to draw one bear from the bag. The color of the bear actually drawn is recorded in two places. All players record the color in the bears at the bottom of the page, and the player who drew the bear makes a record of it by coloring the *Actual* bear opposite the guess for that round. This recording allows each player to deduce the number of bears of each color still in the bag before each turn.
7. Allow time for students to play several rounds, each with a different bag of bears. They will each need one copy of the first student page for every game played.
8. Once students have had several experiences, distribute the second student page. Give each group enough additional bears so that they have 12 of each color.
9. Explain the challenge and give groups time to complete it. Make sure that they know that each group member should start with all 12 bears in the bag before drawing one as a sample.
10. Distribute the final two student pages to allow students to apply their knowledge and analyze their experiences.
11. Close with a time of class discussion and sharing where you go over the questions and have students explain their thinking.

Connecting Learning
1. How did you guess what color of bear you were going to draw? How often were you right?
2. Were you more often correct at the beginning or end of a round? Why do you think this is?
3. For which bags of bears was it easiest to predict the color of the first bear? ...hardest? Why?
4. What do you need to do to make it impossible to draw a green bear? ...certain?
5. Describe a bag from which it would be unlikely to draw a green bear. ...likely to draw a green bear.
6. In the bags your group created, did anyone ever draw a color that was impossible? Why or why not? Did anyone ever draw a color that was unlikely?

Extensions
1. Increase the number of bears and/or the number of colors of bears.
2. Have students describe the probability of pulling each color from the different bags as ratios and fractions. For example, there is a 7:12 chance of drawing a yellow bear from this bag, or the probability of getting green is ½.

* © Copyright 2010. National Governors Association Center for Best Practices and Council of Chief State School Officers. All rights reserved.

Key Question

Which color bear are you most likely to draw from a bag?

Learning Goals

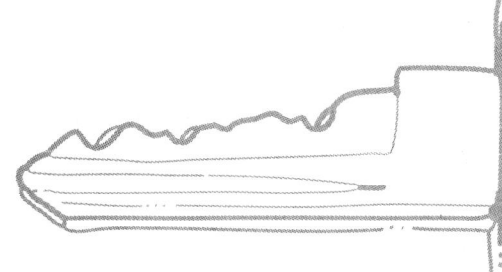

- predict the color of the bear they are likely to draw from a bag,

- draw and record the color of bear they actually draw, and

- evaluate the likelihood of certain events occurring.

#

Our bag has _____ yellow bears and _____ green bears.

Color the number of yellow and green bears you have in your bag.

Guess what color bear you will draw and show what color you did draw.

Guess 🐻   Actual 🐻
Guess 🐻   Actual 🐻
Guess 🐻   Actual 🐻
Guess 🐻   Actual 🐻
Guess 🐻   Actual 🐻
Guess 🐻   Actual 🐻

Color the bears to match what has been picked by you or your partner(s).

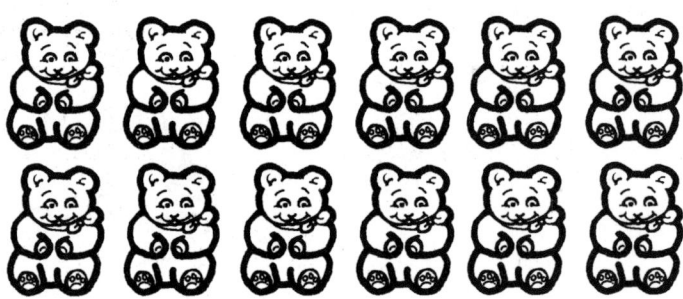

SOLVE IT! 4th — 88 — © 2012 AIMS Education Foundation

Primarily Pro-bear-bility

For each challenge, create a bag of 12 bears. Color the bears to show what you put in the bag. Test your bag by drawing one bear. Record its color.

1. Fill a bag so that it is **impossible** to draw a yellow bear.

   What color bear did you draw? ____

   Show the bears in your bag.

2. Fill a bag so that it is **likely** (not certain) you will draw a green bear.

   What color bear did you draw? ____

   Show the bears in your bag.

3. Fill a bag so that it is **certain** you will draw a yellow bear.

   What color bear did you draw? ____

   Show the bears in your bag.

4. Fill a bag so that it is **unlikely** (not impossible) you will draw a green bear.

   What color bear did you draw? ____

   Show the bears in your bag.

Read the statements. Circle the face that shows the correct response.

1. There are 12 yellow bears and 0 green bears. I drew a yellow bear.

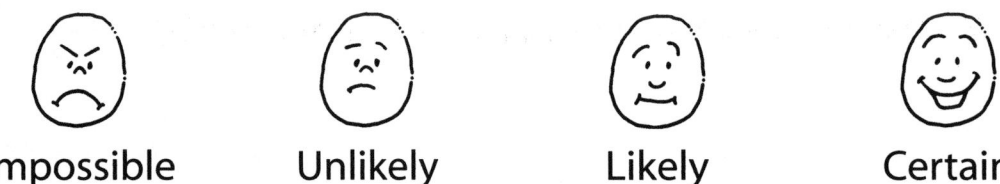

Impossible   Unlikely   Likely   Certain

2. There are 8 green bears and 4 yellow bears. I drew a green bear.

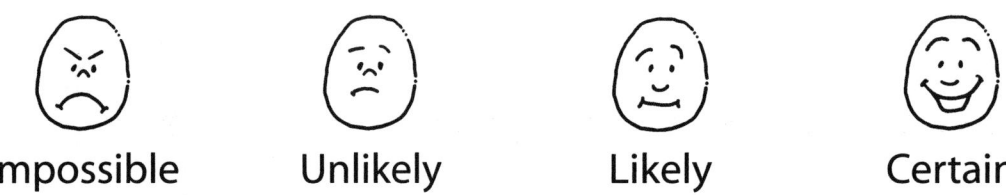

Impossible   Unlikely   Likely   Certain

3. There are 0 yellow bears and 12 green bears. I drew a yellow bear.

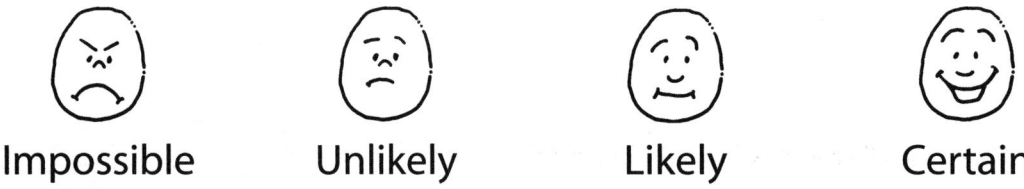

Impossible   Unlikely   Likely   Certain

4. There are 3 green bears and 9 yellow bears. I drew a green bear.

Impossible   Unlikely   Likely   Certain

1. In one round, what was the most yellow bears you picked?

2. How many yellow bears were in your bag that round?

3. Was it more likely for you to draw a green bear or a yellow bear? Why?

4. In one round, what was the most green bears you picked?

5. How many green bears were in your bag that round?

6. Was it more likely for you to draw a green bear or a yellow bear? Why?

7. How often did you correctly guess the color of the bear you drew?

8. Did you have more correct guesses at the beginning of a round or the end of a round? Why?

Connecting Learning

1. How did you guess what color of bear you were going to draw? How often were you right?

2. Were you more often correct at the beginning or end of a round? Why do you think this is?

3. For which bags of bears was it easiest to predict the color of the first bear? …hardest? Why?

4. What do you need to do to make it impossible to draw a green bear? …certain?

5. Describe a bag from which it would be unlikely to draw a green bear. …likely to draw a green bear.

6. In the bags your group created, did anyone ever draw a color that was impossible? Why or why not? Did anyone ever draw a color that was unlikely?

FUEL FILLING FIGURES

Topic
Number combinations

Key Question
Given a certain amount of fuel, how can you fill two or more planes to capacity and use all of your fuel?

Learning Goal
Students will find a way to use a given amount of fuel and fill two or more airplanes in such a way that all of the fuel is used and the planes chosen are filled to capacity.

Guiding Document
*Common Core State Standards for Mathematics**
- Make sense of problems and persevere in solving them. (MP1)
- Solve problems involving measurement and conversion of measurements from a larger unit to a smaller unit. (4.MD)

Math
Number and operations
   addition
      number combinations
      odd/even numbers
Problem solving

Integrated Processes
Observing
Comparing and contrasting
Recording

Problem-Solving Strategies
Work backwards
Guess and check

Materials
Student pages

Background Information
This activity uses airplane fuel capacities to explore number combinations. Given a certain amount of fuel, students are challenged to fill two or more airplanes in such a way that all of the fuel is used and the planes chosen are filled to capacity. All of the planes selected for this activity are of historical significance to the development of airplane flight as we know it today, and a page of plane information is provided to give students some interesting background.

At first, many may use trial and error when looking for solutions, but they should eventually be able to notice a few "rules" that will save them time and help them arrive at answers more quickly. One of these rules has to do with odd and even numbers. When adding, the only way to have an odd total is if one (or any odd number) of the numbers being added is odd. Because only one of the planes—the Curtiss "Jenny"—holds an odd number of gallons of fuel, that plane must be used every time the total number of gallons given is odd. This same reasoning lets you deduce that the Jenny will never be used when the total number of gallons is even. Of course, if you can use the same plane more than once, the Jenny could be used twice (or any even number of times) to achieve an even total amount of fuel.

Management
1. This activity is well suited for individual work, but a time of whole-class discussion to wrap up the experience is vital and should not be overlooked.
2. There are two pages, each with a different challenge. Select the page(s) most appropriate for your students. On the first activity page, students may not use the same plane more than once when selecting the planes to fill. This limits the number of possible combinations and simplifies the process students must go through to arrive at their answers. The second student page presents the same kinds of challenges, but removes the restriction about not using more than one of the same plane. The number of possible combinations is therefore greatly increased, allowing some fuel amounts to be equaled by two, three, and four planes.

Procedure
1. Distribute the student pages to each student and allow them time to complete the problems.
2. Once students have completed both pages, spend some time as a class discussing how students solved the various problems.

Connecting Learning
1. How did you go about solving the problems?
2. What "rules" did you discover as you were solving the problems? (See *Background Information*.)
3. How did these rules help you as you were searching for solutions?
4. Which problems had multiple solutions? Do you think we have found all of the possible answers for each problem?

Extensions
1. Find all of the two- three- and four-plane combinations and how much fuel it would take to fill each. This can be done with or without the restriction about using the same plane more than once in a combination.
2. Look at the fuel capacities of larger planes (the 747-400 has a fuel capacity of 57,285 gallons) and determine the number of smaller planes it would take to equal the fuel capacity of one large plane.
3. Research fuel mileage and determine how far each plane can fly on the amount of fuel it can carry. Do the distances flown correlate to the fuel capacities, or are some airplanes more efficient than others?

Solutions
First Activity Page
1. With 33 gallons of fuel, fill a Piper Cub (12 gallons) and a Curtiss Jenny (21 gallons).
2. With 86 gallons of fuel, fill a Sopwith Camel (30 gallons) and a Cessna Skyhawk (56 gallons).
3. With 75 gallons of fuel, fill a Piper Cub (12 gallons), a Curtiss Jenny (21 gallons), and a Travel Air 3000 (42 gallons).
4. With 93 gallons of fuel, fill a Curtiss Jenny (21 gallons), a Sopwith Camel (30 gallons), and a Travel Air 3000 (42 gallons).
5. With 105 gallons of fuel, fill a Piper Cub (12 gallons), a Curtiss Jenny (21 gallons), a Sopwith Camel (30 gallons), and a Travel Air 3000 (42 gallons).
6. With 63 gallons of fuel, you can fill either a Curtiss Jenny (21 gallons) and a Travel Air 3000 (42 gallons); or a Piper Cub (12 gallons), a Curtiss Jenny (21 gallons), and a Sopwith Camel (30 gallons).
7. All of the two-plane combinations and their total fuel requirements are:
Piper Cub and Curtiss Jenny—33 gallons
Piper Cub and Sopwith Camel—42 gallons
Piper Cub and Travel Air 3000—54 gallons
Piper Cub and Cessna Skyhawk—68 gallons
Curtiss Jenny and Sopwith Camel—51 gallons
Curtiss Jenny and Travel Air 3000—63 gallons
Curtiss Jenny and Cessna Skyhawk—77 gallons
Sopwith Camel and Travel Air 3000—72 gallons
Sopwith Camel and Cessna Skyhawk—86 gallons
Travel Air 3000 and Cessna Skyhawk—98 gallons

Second Activity Page
1. With 60 gallons of fuel, fill either two Sopwith Camels (30 gallons) or five Piper Cubs (12 gallons).
2. With 80 gallons of fuel, fill two Piper Cubs (12 gallons) and one Cessna Skyhawk (56 gallons).
3. With 111 gallons of fuel, fill one Curtiss Jenny (21 gallons), and three Sopwith Camels (30 gallons).
4. With 72 gallons of fuel, fill either a Sopwith Camel (30 gallons) and a Travel Air 3000 (42 gallons), or a Sopwith Camel (30 gallons) and two Curtiss Jennies (21 gallons).
5. With 84 gallons of fuel, fill either two Travel Air 3000s (42 gallons); a Piper Cub (12 gallons), a Sopwith Camel (30 gallons), and a Travel Air 3000 (42 gallons); or a Piper Cub (12 gallons), two Curtiss Jennies (21 gallons), and a Sopwith Camel (30 gallons).

* © Copyright 2010. National Governors Association Center for Best Practices and Council of Chief State School Officers. All rights reserved.

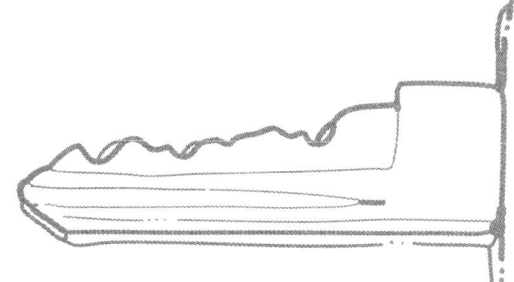

Key Question

Given a certain amount of fuel, how can you fill two or more planes to capacity and use all of your fuel?

Learning Goal

find a way to use a given amount of fuel and fill two or more airplanes in such a way that all of the fuel is used and the planes chosen are filled to capacity.

FUEL FILLING FIGURES
PLANE SPECIFICATIONS

Curtiss JN-4D "Jenny"
First produced: 1916
Fuel capacity: 21 gallons
Facts: This biplane was very popular during World War I. It was used to train pilots. About 90% of North American pilots during this time learned to fly in a Jenny.

Sopwith "Camel"
First produced: 1917
Fuel capacity: 30 gallons
Facts: The Camel was a fighter plane in World War I. It was very difficult for inexperienced pilots to fly. Roy Brown was flying in a Camel when he shot down "The Red Baron."

Travel Air 3000
First produced: 1927
Fuel capacity: 42 gallons
Facts: Travel Air biplanes were very popular for racing pilots. They were also popular for endurance flights. Charles Lindbergh asked Travel Air to build him a plane for his historic flight across the Atlantic Ocean, but they were too busy with other orders.

Piper J-3 "Cub"
First produced: 1938
Fuel capacity: 12 gallons
Facts: This plane is small and easy to fly. It was very popular during World War II, when as many as 75% of American pilots used it to learn to fly.

Cessna Skyhawk 172
First produced: 1955
Fuel capacity: 56 gallons
Facts: The Cessna 172 is small, single-engine plane that is very popular today. Most flight schools use it to train pilots. There have been more than 35,000 C172s produced since 1955.

FUEL FILLING FIGURES

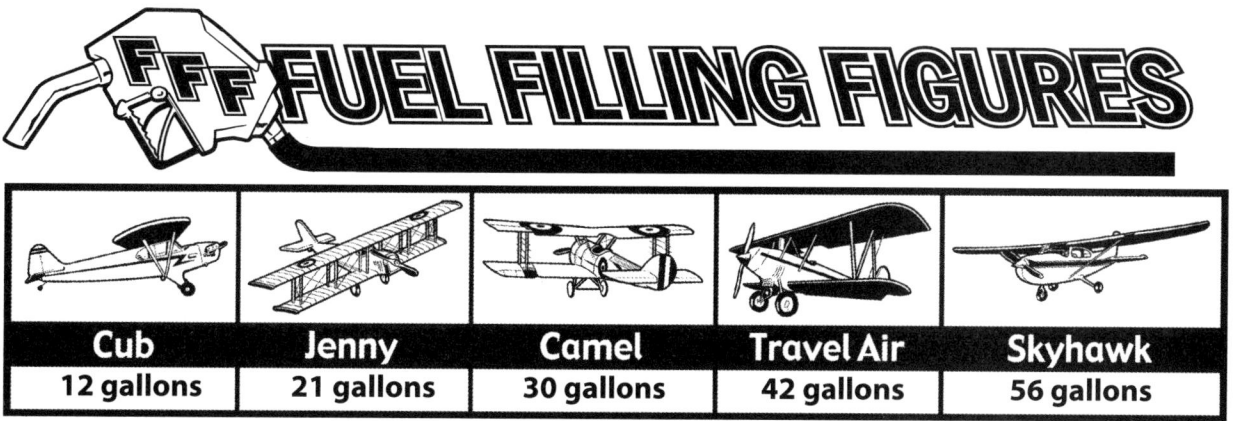

Cub	Jenny	Camel	Travel Air	Skyhawk
12 gallons	21 gallons	30 gallons	42 gallons	56 gallons

Solve each of the problems below using the plane information you have been given. In each case, you must fill the gas tanks of two or more planes. Each tank must be filled to capacity, and there can be no extra fuel left. You may not use more than one of the same plane.

1. You have 33 gallons of fuel. Which planes will you fill?

2. You have 86 gallons of fuel. Which planes will you fill?

3. You have 75 gallons of fuel. Which planes will you fill?

4. You have 93 gallons of fuel. Which planes will you fill?

5. You have 105 gallons of fuel. Which planes will you fill?

6. You have 63 gallons of fuel. How can you fill just two planes? How can you fill three planes?

7. List all of the possible two-plane combinations. How much fuel would you need to fill both planes in each case?

SOLVE IT! 4th 97 © 2012 AIMS Education Foundation

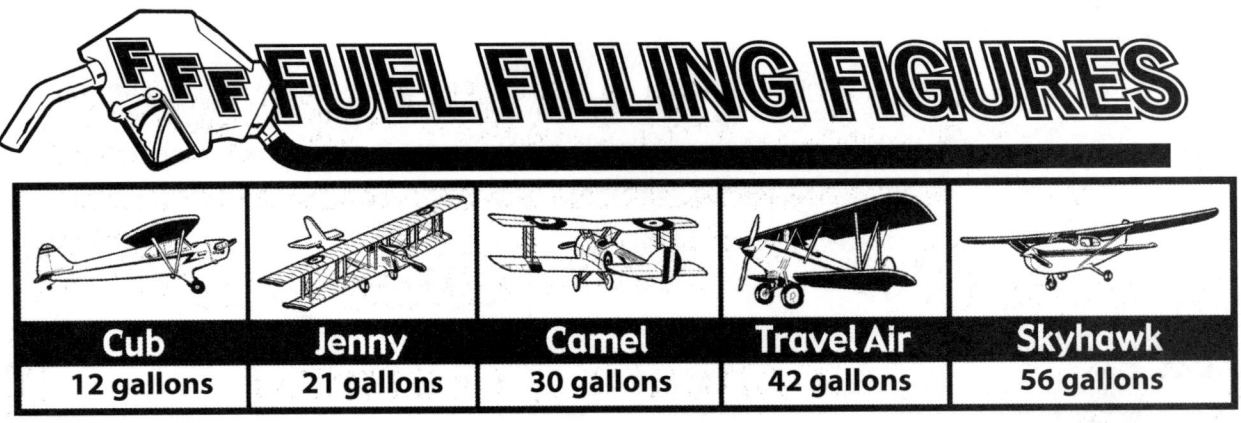

Cub	Jenny	Camel	Travel Air	Skyhawk
12 gallons	21 gallons	30 gallons	42 gallons	56 gallons

Solve each of the problems below using the plane information you have been given. In each case, you must fill the gas tanks of two or more planes. Each tank must be filled to capacity, and there can be no extra fuel left. This time, you may use more than one of the same plane.

1. You have 60 gallons of fuel. Which planes will you fill?

2. You have 80 gallons of fuel. Which planes will you fill?

3. You have 111 gallons of fuel. Which planes will you fill?

4. You have 72 gallons of fuel. How can you fill exactly two planes? How can you fill exactly three planes?

5. You have 84 gallons of fuel. How many ways can you fill exactly two, exactly three, and exactly four planes?

Use the back of this paper to create your own fuel filling problems.

SOLVE IT! 4th   © 2012 AIMS Education Foundation

FUEL FILLING FIGURES

Connecting Learning

1. How did you go about solving the problems?

2. What "rules" did you discover as you were solving the problems?

3. How did these rules help you as you were searching for solutions?

4. Which problems had multiple solutions? Do you think we have found all of the possible answers for each problem?

Mix-Ups and Mysteries

Topic
Problem solving

Key Question
How can you work backwards from the information given to find the money mix-up and the missing muffins?

Learning Goals
Students will:
- read a series of word problems dealing with money and elapsed time, and
- work backwards to solve the problems posed.

Guiding Document
Common Core State Standards for Mathematics*
- *Make sense of problems and persevere in solving them. (MP1)*
- *Solve problems involving measurement and conversion of measurements from a larger unit to a smaller unit. (4.MD)*

Math
Money
Elapsed time
Problem solving

Integrated Processes
Observing
Comparing and contrasting
Recording

Problem-Solving Strategies
Work backwards
Use manipulatives

Materials
Paper fasteners
Scissors
Student pages
#19 rubber bands
Paper plates, optional
Glue sticks, optional

Background Information
We often use the skill of working backwards in our daily lives, particularly when dealing with money and elapsed time. We calculate when to set our alarm based on how long it takes us to get ready and when we need to leave. We determine how much money to take on vacation based on the items we expect to buy and how much they cost. The word problems in this activity give students the opportunity to practice these skills in a problem-solving context.

Management
1. There are two scenarios presented in this activity. The first deals with money and is fairly simple. The second deals with elapsed time and is more difficult. You may decide to use only one or both of these, depending on the needs and abilities of your students.
2. For the elapsed time scenario, students will need to construct a clock to assist them as they work through the problems. They may simply cut out the clock hands and fasten them to the face provided, or the face may also be cut out and mounted to a paper plate to make it sturdier. Either way, it is recommended that you copy the clock page onto card stock.
3. To assemble the rubber band books, fold the pages in half horizontally and then vertically. Nest them together so that the pages are in order and hold them together with a #19 rubber band.

Procedure
1. Distribute the pages for *The Mall Money Mix-Up* rubber band book. Have students assemble the books.
2. Provide time for students to read the books and answer the questions. If desired, students may work in small groups.
3. Discuss students' answers and how they came up with those answers.
4. If appropriate, distribute the materials for making the clocks and *The Mystery of the Missing Muffins* rubber band book.
5. Allow students to work together in groups to try and determine the muffin thief.
6. Discuss their solutions, how they got those solutions, and how this problem compared to the previous one.

SOLVE IT! 4th © 2012 AIMS Education Foundation

Connecting Learning

1. How much money did the girls start with? [$25.00]
2. Which sister is mixed up with her math? [Fluffy]
3. How much money did she really have left at the end of the day? [$3.00] How do you know?
4. Who was the muffin thief? [Sam Shoemaker] How do you know? [15 minutes of his day are unaccounted for]
5. How did you go about solving these problems?
6. Did you use the clock to help you? Why or why not?
7. Was it easier to figure out the money problem or the time problem? Why?
8. What are some times you have used working backwards to solve a real-life problem? [deciding when to get up, knowing when to leave for a movie, figuring out how much money you will have left after a purchase, etc.]

Extension

Have students create their own money or time riddles and share them with classmates.

* © Copyright 2010. National Governors Association Center for Best Practices and Council of Chief State School Officers. All rights reserved.

Mix-Ups and Mysteries

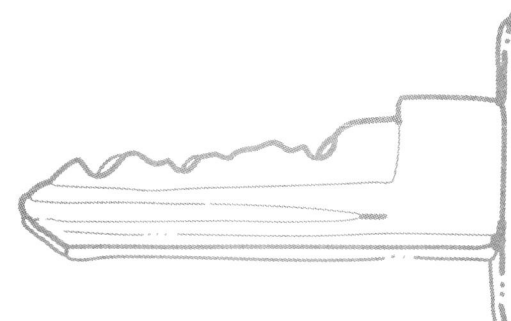

Key Question

How can you work backwards from the information given to find the money mix-up and the missing muffins?

Learning Goals

Students will:

- read a series of word problems dealing with money and elapsed time, and

- work backwards to solve the problems posed.

The Mall Money Mix-Up

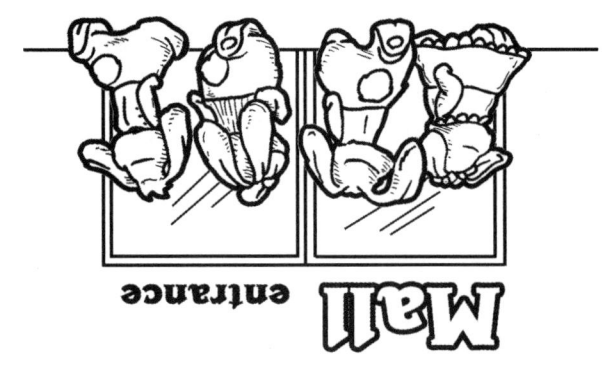

Four sisters went to the mall. They all started the day with the same amount of money. They all bought different things. One is a little mixed up with her math. She says she has more money left than is possible. Find out who is really ending with less money than she thinks.

How much money did the girls start with?

Which sister is mixed up with her math?

How much money did she really have left at the end of the day? Show your work.

Buffy: There was a great sale at CD Cellout. Buy three; get one free! The CDs were $7.00 each. I got four great albums by Richard Rocker. It was dollar day at Ice Cream World. Each scoop is only a buck! So I got rocky road and pistachio. I had $2.00 left at the end of the day.

Fluffy: I needed new earrings. They were on sale three pairs for $12.00. I got gold hoops, silver stars, and some that look like diamonds. Then I found a great necklace for only $9.00. I bought a single scoop of ice cream at Ice Cream World and still had $5.00 left over.

Muffy: My new scarf is fabulous! It was only $8.00 when I bought the matching hat for $14.00. I had to buy the gloves too—they were only $3.00! But I didn't have any money left to eat ice cream with the other girls.

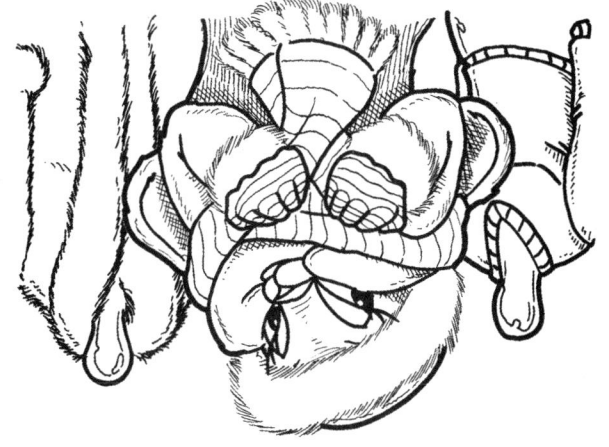

Huffy: The things I like are never on sale! There was a beautiful sweater at Wool World, but it was $30.00, and I didn't have enough money. Muffy bought the last hat, so my scarf cost $12.00. I did find some fun socks for $5.00. To make up for feeling bad, I had three scoops at Ice Cream World. I'll save my extra $5.00 for next month's shopping trip.

The Mystery of the Missing Muffins

Betty Baker is missing some muffins. Sometime this morning, a thief stole eight muffins from her shop. Four suspects gave their stories to the police. Help solve the crime by finding whose story doesn't add up. Someone is leaving out part of his or her story—and a few minutes of time, too. If you can find the missing minutes, you will know the identity of the muffin thief. Good luck!

Tricia Teacher: I woke up early this morning because I had so much to do. I was at school an hour before it starts. I graded for 30 minutes. Then I talked to Mrs. Jimenez next door for 10 minutes. I went to the office to make copies at 7:55. That took another 10 minutes. When I got back to my room, I remembered that I was supposed to get muffins for the teacher break. But by then it was too late. School was starting in less than 10 minutes.

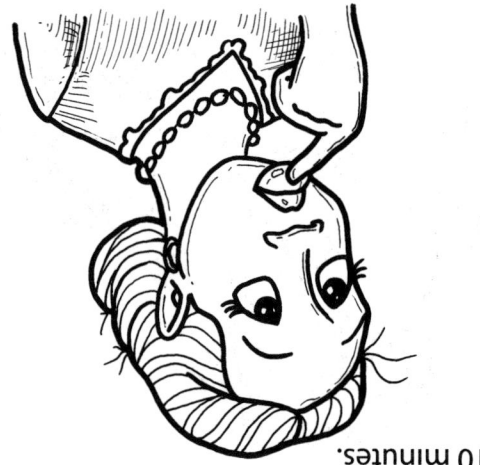

Who stole the muffins? Explain how you solved the problem below.

Betty Baker: I got to work at 5:30 A.M. as usual. By 6:00, I had two dozen hot blueberry muffins out on the counter. At 6:10, Chief Charlie came in and bought a dozen for a meeting. Then there were a few of my usual customers coming by to talk and eat. I put out another two dozen muffins at 7:00. At 7:45, I always count what I have. There were eight muffins left. A few minutes later, I'm not sure when exactly, there was a large crash in the back of the store. I rushed to see what had happened. My large metal mixing bowl had fallen off the counter. Then the phone rang. It was my blueberry supplier, and we talked for a while. When I came back out front, all of the muffins were gone! I called Chief Charlie, and he was there by 8:00 to start the investigation.

Lenny Lawyer: I got to work today at 8:00 sharp. I usually stop at Betty's Bakery on the way in, but today I was running late. It takes me 20 minutes to get to work. It takes 30 minutes if I stop at the bakery. My alarm goes off at 6:30. I overslept by 45 minutes this morning. I got ready in only 25 minutes. I barely made it out the door on time. There's no way I could have stopped at the bakery.

Sam Shoemaker: My shoe shop is next to Betty's Bakery. It opens at 8:30, but I'm always there half an hour early. Today, I got up at 6:30 and spent 30 minutes reading the paper and eating breakfast. Then I took a 10 minute shower. Only 15 minutes after that, I was out the door. Since I live less than two miles away, I have a nice 20 minute walk to work.

Sally Schoolgirl: School starts at 8:15. My teacher is Mrs. Jimenez. Today I was there 10 minutes early. It takes me 20 minutes to walk to school. To get ready, I need 45 minutes. This morning I woke up at 7:00. The bakery is 10 minutes from school. I walk by it every day, but I never have any money, so I don't stop.

Mix-Ups and Mysteries

Cut out the clock hands. Attach them to the clock face with a paper fastener. If desired, mount the clock face on a paper plate.

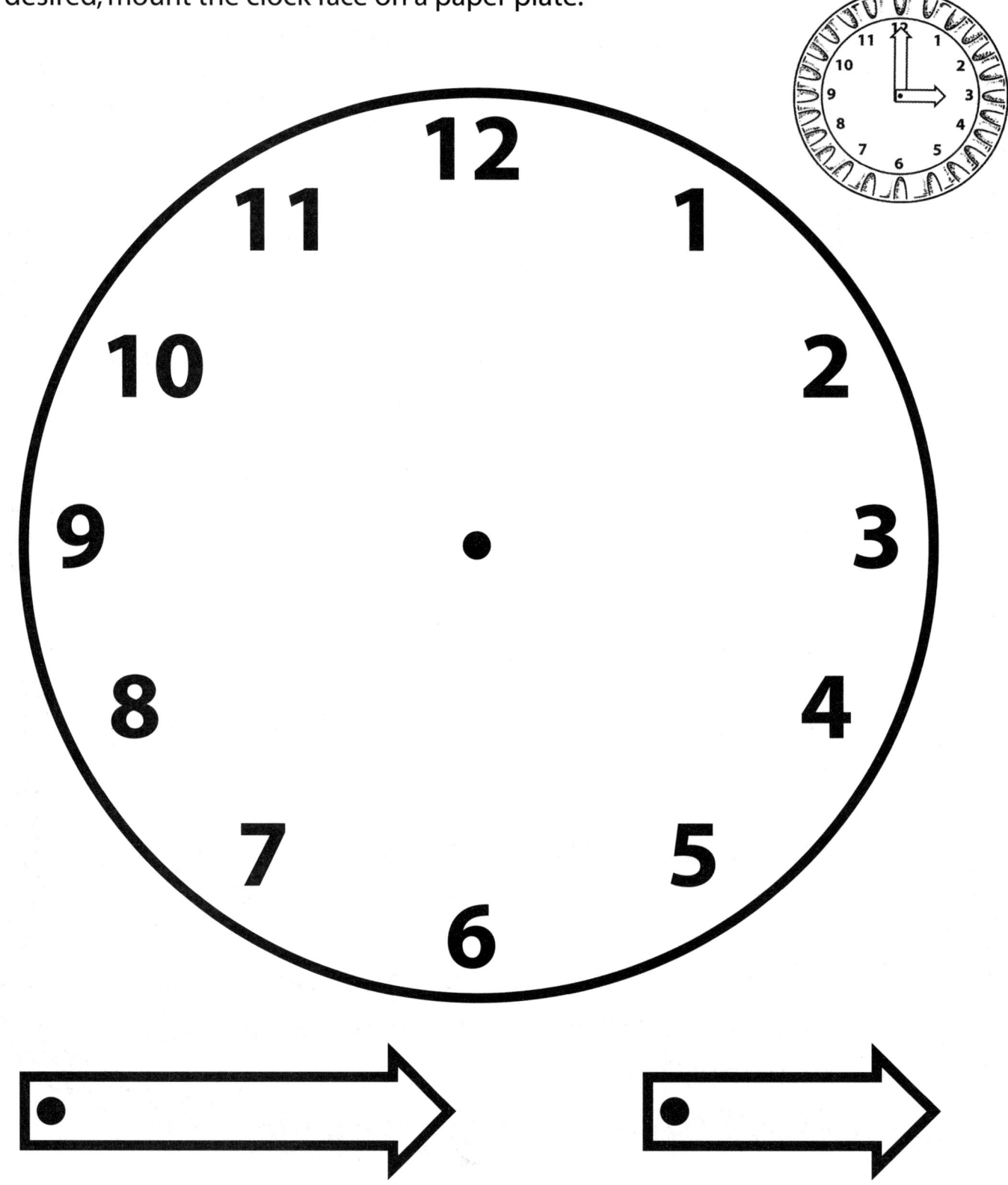

Mix-Ups and Mysteries

Connecting Learning

1. How much money did the girls start with?

2. Which sister is mixed up with her math?

3. How much money did she really have left at the end of the day? How do you know?

4. Who was the muffin thief? How do you know?

5. How did you go about solving these problems?

6. Did you use the clock to help you? Why or why not?

7. Was it easier to figure out the money problem or the time problem? Why?

8. What are some times you have used working backwards to solve a real-life problem?

Problem-Solving Strategies
Organize the Information

It is often helpful to organize the information when trying to solve problems. You can put what you know into a list, chart, or table. Then you can see what you still need to solve the problem. You can also use this strategy when you need to find lots of different solutions.

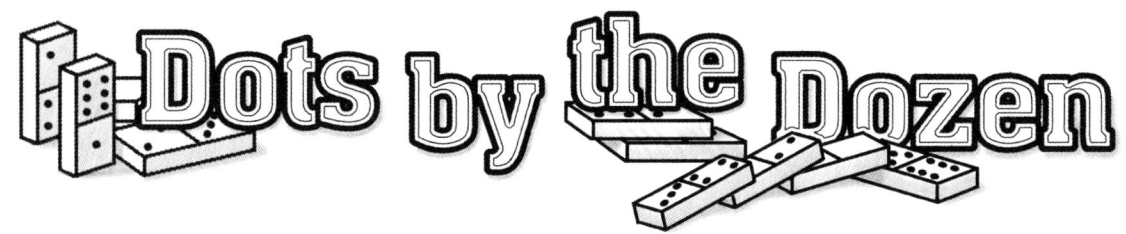

Dots by the Dozen

Topic
Data organization

Key Question
What are all of the two-domino combinations that have a total of 12 dots?

Learning Goals
Students will:
- find every two-domino combination that has a total of 12 dots,
- organize those solutions in a logical way, and
- extend the problem to find 10- and 14-dot combinations.

Guiding Document
Common Core State Standards for Mathematics*
- *Make sense of problems and persevere in solving them. (MP1)*
- *Look for and make use of structure. (MP7)*

Math
Data organization
Number combinations
Problem solving

Integrated Processes
Observing
Comparing and contrasting
Organizing
Recording

Problem-Solving Strategies
Organize the information
Look for patterns
Use manipulatives

Materials
Double-six dominoes, one set per group
Student pages

Background Information
This activity takes the familiar double-six dominoes and uses them to create a challenging problem that requires organization, careful analysis of solutions, and mathematical communication. The challenge is to find every two-domino combination that has a total of 12 dots, organize those solutions in a logical way, and then extend the problem to find 10- and 14-dot combinations.

As mentioned, the initial activity asks students to find every two-domino combination that has a total of 12 dots. (Double-six and double-blank, six-four and double-one, three-two and five-two, etc.) Because they will be working with a single set of dominoes, each solution must use two different dominoes. For example, even though two double-three dominoes would total 12 dots, that solution is not possible given the constraints of the problem.

Management
1. This activity requires the use of a standard set of double-six dominoes. Students should work together in pairs or groups of three, and each group will need its own set of dominoes. If you do not have enough sets of real dominoes, you can make copies of the domino pages included with this activity.
2. Because there are multiple parts to this activity, it is recommended that you spread it out over the course of several days to a week.
3. Students will need two copies each of the solutions recording page.

Procedure
1. Have students get into groups and distribute the dominoes and student pages.
2. Allow time over several days for students to work on the problem and discover solutions.
3. Once the problem has been thoroughly explored, spend some time going through the questions on the first student sheet.
4. As a class, explore several different ways to organize the solutions and discuss the pros and cons of each. Talk about the different patterns that are revealed by the different organizational schemes and how some organizations may help you to see if all of the solutions have been discovered.
5. After the initial problem has been fully explored, challenge students to find two-domino combinations that total 10 and 14 dots. If desired, you may have half of the groups work on the 10-dot problem while the other half works on the 14-dot problem.

6. Once students have found all of the solutions, compare and contrast the three problems. Look at the total number of solutions for each, which dominoes were used most frequently, any patterns in the solutions, etc.

Connecting Learning
1. How many solutions did you find for 12 dots? Do you think you have found them all? Why or why not?
2. Which dominoes are used the most often in the solutions? ...the least often? Why do you think this is?
3. What is the most number of times any single domino is used in different solutions? What is the least number of times any single domino is used?
4. How did you choose to organize your solutions?
5. How does this method compare to the methods used by other groups? Which do you think is most effective? Why?
6. How many solutions did you find for 10 dots? ...14 dots? How does this compare to the number of solutions for 12 dots?
7. What patterns do you see the solutions?

Extension
Have students find all two-domino combinations that have totals from nine to 15. Some interesting patterns in the total numbers of solutions can be seen when you look at several consecutive sums.

Solutions

The solutions shown here are organized in two different ways. The solutions for 12 dots are organized by the two-number combinations that sum to 12 (12 + 0, 11 + 1, etc.). The solutions for 10 and 14 dots are organized by the numbers they contain (solutions with ones, solutions with twos, etc.).

Solutions for 12 dots

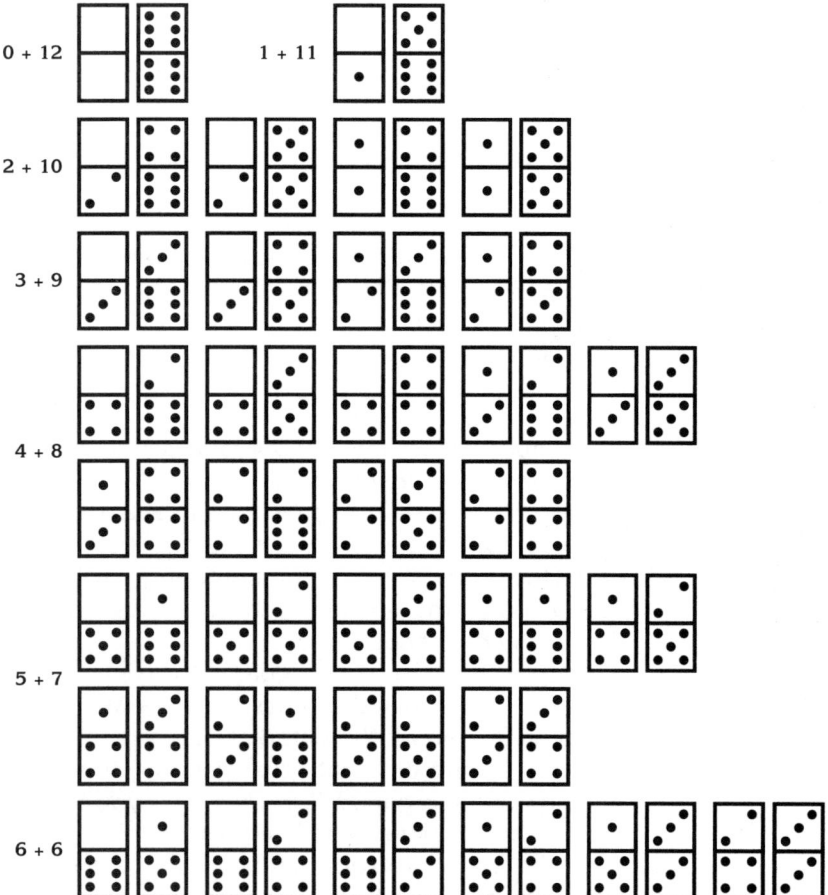

SOLVE IT! 4th   114   © 2012 AIMS Education Foundation

Solutions for 10 dots

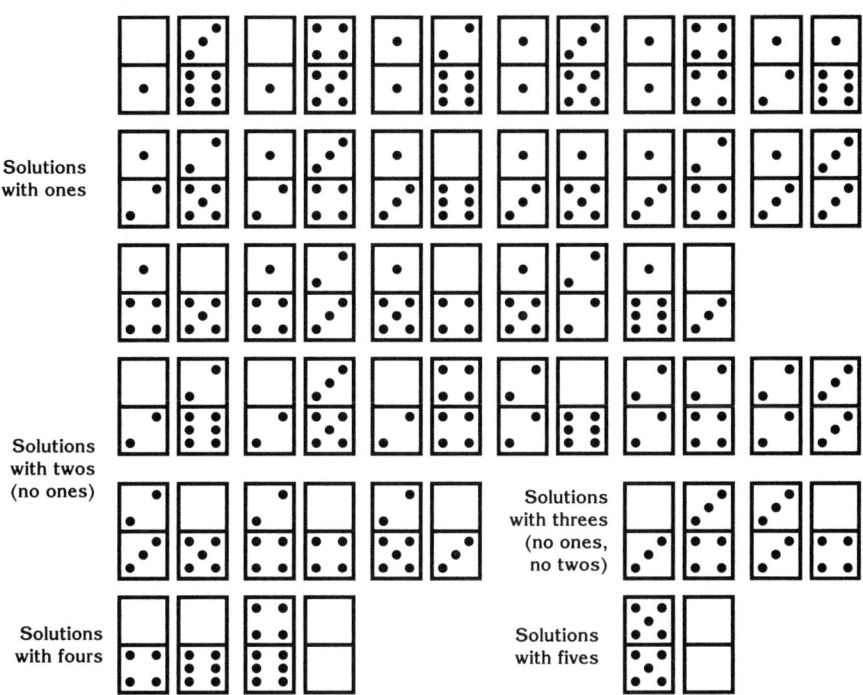

Solutions with ones

Solutions with twos (no ones)

Solutions with threes (no ones, no twos)

Solutions with fours

Solutions with fives

Solutions for 14 dots

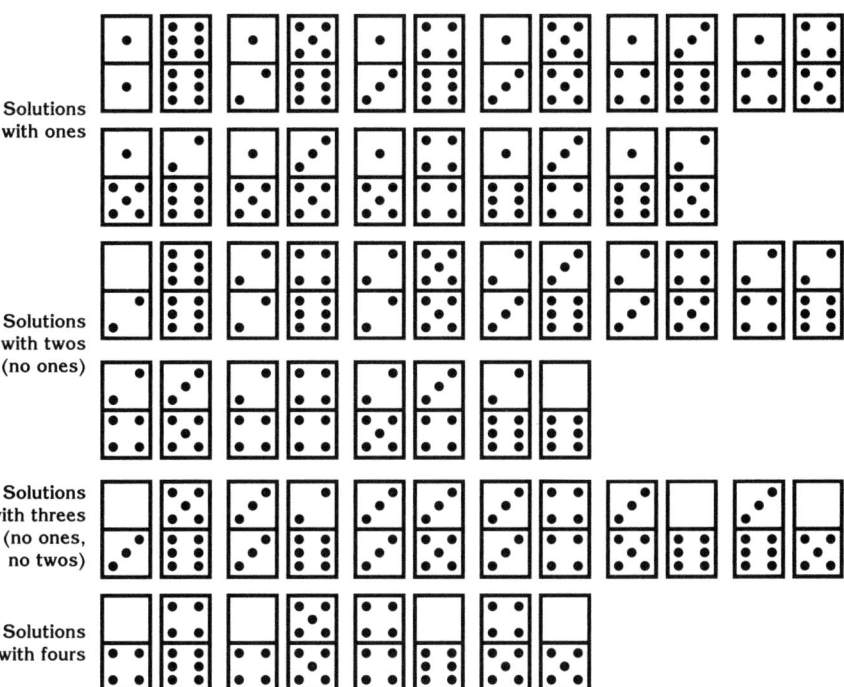

Solutions with ones

Solutions with twos (no ones)

Solutions with threes (no ones, no twos)

Solutions with fours

* © Copyright 2010. National Governors Association Center for Best Practices and Council of Chief State School Officers. All rights reserved.

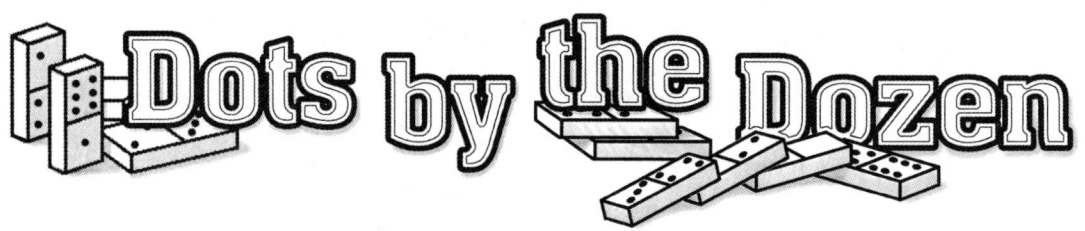

Dots by the Dozen

Key Question

What are all of the two-domino combinations that have a total of 12 dots?

Learning Goals

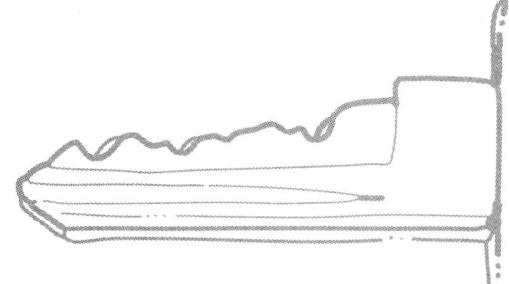

- find every two-domino combination that has a total of 12 dots,

- organize those solutions in a logical way, and

- extend the problem to find 10- and 14-dot combinations.

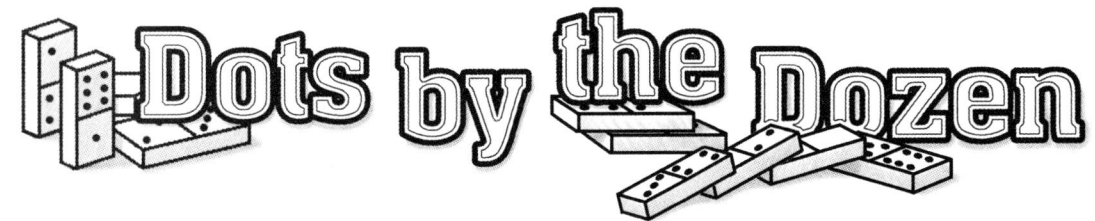

Dots by the Dozen

Find every combination of two dominoes that has a total of 12 dots. Record your solutions on the next page and then answer the questions below. You may not use the same domino twice in a single solution. (For example, you cannot use two double-three dominoes even though they would total 12.)

1. How many solutions did you find? Do you think you have found them all? Why or why not?

2. Which dominoes are used the most often in the solutions? …the least often? Explain why this is the case.

3. What is the most number of times any single domino is used in different solutions? What is the least number of times any single domino is used?

4. On a second copy of the solutions page, organize your solutions in a logical fashion. Describe your organizational scheme below.

5. If you repeated this problem finding combinations of dominoes that have a total of 10 dots, do you think you would find more solutions, fewer solutions, or the same number of solutions? How about 14 dots? Justify your responses.

SOLVE IT! 4th © 2012 AIMS Education Foundation

Record each solution you discover in the spaces below.

SOLVE IT! 4th

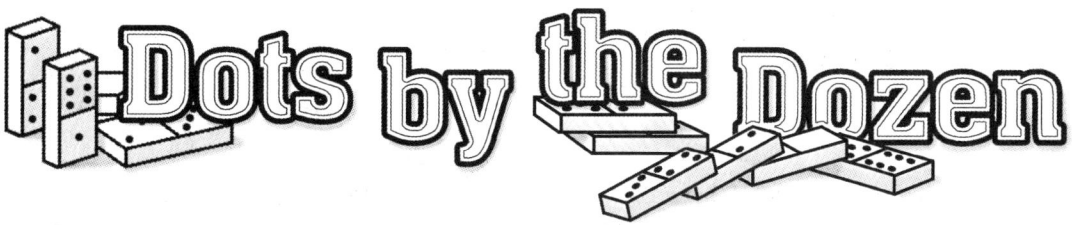

Dots by the Dozen

Connecting Learning

1. How many solutions did you find for 12 dots? Do you think you have found them all? Why or why not?

2. Which dominoes are used the most often in the solutions? …the least often? Why do you think this is?

3. What is the most number of times any single domino is used in different solutions? What is the least number of times any single domino is used?

4. How did you choose to organize your solutions?

5. How does this method compare to the methods used by other groups? Which do you think is most effective? Why?

6. How many solutions did you find for 10 dots? …14 dots? How does this compare to the number of solutions for 12 dots?

7. What patterns do you see the solutions?

That's Sum Pattern!

Topic
Patterns

Key Question
What are all the pairs of whole numbers that have a sum of 15?

Learning Goals
Students will:
- find all of the two-number combinations that total 15, and
- organize their answers to be sure they have found them all.

Guiding Documents
Project 2061 Benchmark
- *Mathematics is the study of many kinds of patterns, including numbers and shapes and operations on them. Sometimes patterns are studied because they help to explain how the world works or how to solve practical problems, sometimes because they are interesting in themselves.*

Common Core State Standards for Mathematics*
- *Make sense of problems and persevere in solving them. (MP1)*
- *Look for and make use of structure. (MP7)*

Math
Pattern recognition
Data organization
Number and operations
 addition
Problem solving

Integrated Processes
Observing
Recording
Organizing
Comparing and contrasting

Problem-Solving Strategies
Organize the information
Look for patterns
Guess and check

Materials
Student page

Background Information
That's Sum Pattern! is designed to highlight several problem-solving strategies. One of the most important of these strategies, and a key problem-solving skill that all students need to learn, is *organizing data*. When data are organized, any patterns that are embedded in the data become more easily recognizable. And, using patterns is yet another important problem-solving strategy.

In addition to highlighting problem-solving strategies, this activity is designed to build students' mathematical vocabulary. As a teacher, you should strive to use the correct mathematical terms whenever possible. For example, instead of asking students to find the answer to an addition problem, ask them to compute the sum. Instead of asking students what they get when they multiply seven times eight, ask them to calculate the product of seven and eight. Instead of telling students to take away 27 from 93, ask them to calculate the difference between 93 and 27.

In the end, students should see, first hand, the power of organizing data and applying this problem-solving skill. The activity ends with a challenge to find all the whole number pairs that have a product of 24. Hopefully, students will not approach this problem randomly, but will start by organizing the data to come up with an organized list or table of all the whole number pairs.

Management
1. This activity should be done in groups, with three to four students working together.
2. Although the commutative property of addition says that two problems with the same addends (e.g., 7 + 8 and 8 + 7) are the same, to see all the patterns that are embedded in this activity, they are considered to be different.
3. In this activity, students will need to know that whole numbers are the natural or counting numbers (1, 2, 3, 4, 5, ...) with the addition of zero (i.e., 0, 1, 2, 3, 4, ...).

Procedure
1. Have students get into groups and distribute the student page.
2. Allow time for groups to search for answers and organize their solutions. Be sure they understand that all combinations, even those that use the same numbers in a different order, should be included.

SOLVE IT! 4ᵗʰ 121 © 2012 AIMS Education Foundation

3. Discuss the methods students used to find all of the solutions and organize them in a logical fashion.
4. Challenge students to use the patterns they discovered and the organizational methods they used to find all of the two-number combinations that sum to 24.
5. Discuss their findings.

Connecting Learning
1. How many pairs of whole numbers have a sum of 15? [16]
2. How do you know that you have found them all?
3. Is the method that you used to organize the same as or different from the methods used by other groups?
4. Do you think one method is more effective than others? Explain your thinking.
5. If solutions that use the same pairs of numbers in a different order are counted as one, how many solutions would there be? [eight]
6. How many pairs of whole numbers have a product of 24? [eight] How do you know?

Extensions
1. Have students write their number pairs as equations (e.g., 2 + 13 = 15). If appropriate, have them recognize the algebraic equation that is the basis of this activity—x + y = 15, where x and y are whole numbers.
2. If your students are familiar with coordinate graphs, have them plot all the number pairs in this activity in the first quadrant of the Cartesian plane.

Solutions
There are 16 pairs of whole numbers that have a sum of 15. This assumes that the same numbers in a different order constitute a unique solution.

0, 15	4, 11	8, 7	12, 3
1, 14	5, 10	9, 6	13, 2
2, 13	6, 9	10, 5	14, 1
3, 12	7, 8	11, 4	15, 0

There are eight pairs of whole numbers that have a product of 24. This assumes that the same numbers in a different order constitute a unique solution.

1, 24	6, 4
2, 12	8, 3
3, 8	12, 2
4, 6	24, 1

* © Copyright 2010. National Governors Association Center for Best Practices and Council of Chief State School Officers. All rights reserved.

THAT'S SUM PATTERN!

Key Question

What are all the pairs of whole numbers that have a sum of 15?

Learning Goals

Students will:

- find all of the two-number combinations that total 15, and

- organize their answers to be sure they have found them all.

THAT'S SUM PATTERN!

What two whole numbers (0, 1, 2, 3, 4, . . .) have a sum of 15? List as many pairs of whole number as you can that have a sum of 15.

Do you have all the pairs? How do you know?

One of the key tools in problem solving is to *organize* data. How can you organize the pairs of numbers above? Think of some ways and record them here.

Pick one way to organize the data. Make an organized list showing all the whole number pairs with a sum of 15.

Challenge:
Use what you learned doing this problem to find all the pairs of whole numbers whose product is 24. Show your work on the back.

THAT'S SUM PATTERN!

Connecting Learning

1. How many pairs of whole numbers have a sum of 15?

2. How do you know that you have found them all?

3. Is the method that you used to organize the same as or different from the methods used by other groups?

4. Do you think one method is more effective than others? Explain your thinking.

5. If solutions that use the same pairs of numbers in a different order are counted as one, how many solutions would there be?

6. How many pairs of whole numbers have a product of 24? How do you know?

Patty's Penny Puzzler

Topic
Problem solving

Key Question
What are all of the possible ways to make 36 cents with exactly one penny and exactly six pennies?

Learning Goal
Students will find as many ways (using any combination of quarters, dimes, nickels, and pennies) as they can to make 36 cents if there is exactly one penny or exactly six pennies.

Guiding Document
*Common Core State Standards for Mathematics**
- *Make sense of problems and persevere in solving them. (MP1)*

Math
Number combinations
Problem solving

Integrated Processes
Observing
Recording
Organizing
Comparing and contrasting

Problem-Solving Strategies
Organize the information
Look for patterns
Guess and check

Materials
Student pages
Coins, optional

Background Information
This activity has multiple solutions. In it, students are challenged to find as many ways (using any combination of quarters, dimes, nickels, and pennies) as they can to make 36 cents if there is exactly one penny (six ways) or exactly six pennies (five ways). Students will use a variety of problem-solving strategies as they work together to find all of the possible solutions.

Management
1. *Patty's Penny Puzzler* can be done in either an open-ended or a more structured way. For the open-ended approach, give the students only the first page and let them work on the problem with a minimum amount of guidance. For a more structured approach, give students the second page in addition to the first page and help them as much as is necessary.
2. If desired, you can provide students with coins (real or play) to manipulate as they try to find all of the possible combinations.

Procedure
1. Distribute the student page(s) and have students get into groups.
2. Go over the problem and be sure everyone understands the task.
3. Allow time for groups to discover as many solutions as they can. If desired, provide coins for them to work with.
4. Close with a time of discussion where groups share their solutions and the methods they used to find them.

Connecting Learning
1. What coins could be in Patty's pocket if she has 36 cents and exactly one penny? (See *Solutions*.)
2. What if she has exactly six pennies? (See *Solutions*.)
3. How did your group find all of the different combinations?
4. What did you do to make sure you hadn't missed any?

Extension
Find all of the ways to make 36 cents using any number of pennies.

Solutions
There are six ways to make 36 cents with one penny and five ways to make 36 cents with six pennies.

One-Penny Combinations
1 quarter, 1 dime, 1 penny
1 quarter, 2 nickels, 1 penny
3 dimes, 1 nickel, 1 penny
2 dimes, 3 nickels, 1 penny
1 dime, 5 nickels, 1 penny
7 nickels, 1 penny

Six-Penny Combinations
1 quarter, 1 nickel, 6 pennies
3 dimes, 6 pennies
2 dimes, 2 nickels, 6 pennies
1 dime, 4 nickels, 6 pennies
6 nickels, 6 pennies

* © Copyright 2010. National Governors Association Center for Best Practices and Council of Chief State School Officers. All rights reserved.

Patty's Penny Puzzler

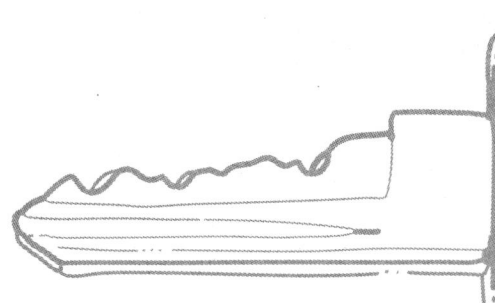

Key Question

What are all of the possible ways to make 36 cents with exactly one penny and exactly six pennies?

Learning Goal

Students will:

find as many ways (using any combination of quarters, dimes, nickels, and pennies) as they can to make 36 cents if there is exactly one penny or exactly six pennies.

Patty's Penny Puzzler

Patty persists in puzzling people with perplexing problems. She has 36¢ and pesters her pals to predict what coins might be in her pocket if she has exactly one penny or exactly six pennies.

List as many of the possible coin combinations as you can.

Do you think you found them all?
Justify your answer.

Explain how you might organize your answers so that the chances of missing any coin combinations are minimized.

SOLVE IT! 4th © 2012 AIMS Education Foundation

Patty's Penny Puzzler

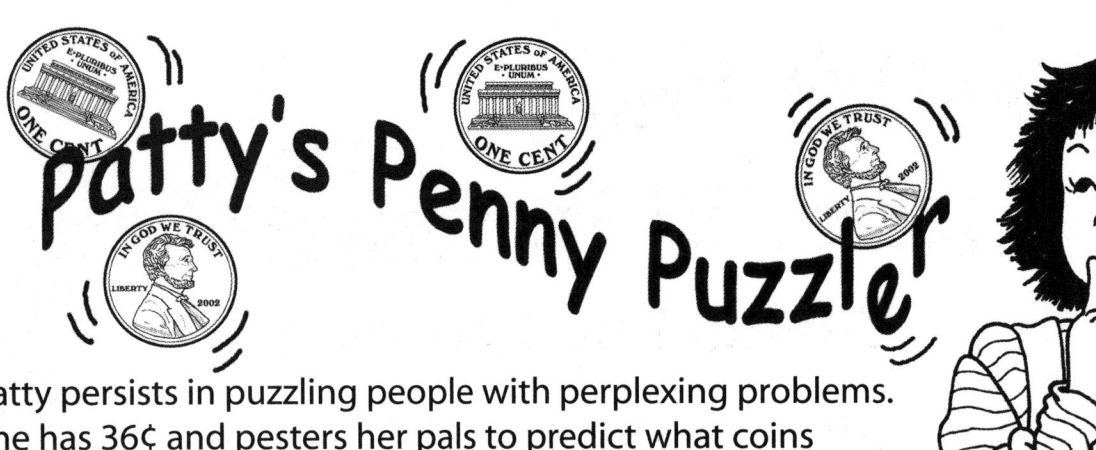

Patty persists in puzzling people with perplexing problems. She has 36¢ and pesters her pals to predict what coins might be in her pocket if she has exactly one penny or exactly six pennies. Use the table below to help you find all the possible coin combinations.

Quarters	Dimes	Nickels	Pennies

SOLVE IT! 4th © 2012 AIMS Education Foundation

Connecting Learning

1. What coins could be in Patty's pocket if she has 36 cents and exactly one penny?

2. What if she has exactly six pennies?

3. How did your group find all of the different combinations?

4. What did you do to make sure you hadn't missed any?

Centennial Celebration Calculations

Topic
Problem solving

Key Question
Which store provides the best deal for the numerals you need to buy for your celebration?

Learning Goals
Students will:
- determine the number of cut-out numerals that need to purchased to make all of the numbers from one to 100, and
- determine which of two stores offers a better deal for the quantity of numerals needed.

Guiding Document
*Common Core State Standards for Mathematics**
- Make sense of problems and persevere in solving them. (MP1)
- Solve problems involving measurement and conversion of measurements from a larger unit to a smaller unit. (4.MD)

Math
Problem solving

Integrated Processes
Observing
Collecting and recording data
Organizing data
Applying

Problem-Solving Strategies
Organize the information
Draw out the problem

Materials
Student page

Background Information
The scenario behind this problem is a hypothetical centennial celebration for the hundredth day of school (which usually falls in February for schools on a traditional schedule). The storyline tells students that their class is in charge of the cafeteria decorations for the celebration. As part of the decoration plans, the class decides to buy fancy cut-out numerals and use them to build all the numbers from one to 100 around the tops of the cafeteria walls. The budget for this project is 20 dollars. After doing some research, students find there are two different pricing options for the numerals. The local discount store has gold numerals for five cents each. The local stationery store has the exact same numerals, but a different pricing scheme: the charge is 10 cents per numeral for the first 10; the next 100 numerals are five cents each; and any numerals beyond that are three cents each. The problem challenges students to find the best deal.

Management
1. This activity can be done using either an open-ended or a structured approach. Use the first student page for an open-ended approach. This page simply presents the problem to students and allows them to solve it in any way they can. The second student page is designed for those who would like a more structured approach. This page gives students more guidance, using tables to help them see the steps they need to go through to solve the problem. Use the page that is most appropriate for your students.
2. Have students work together in small groups.

Procedure
1. Distribute the appropriate student page to your class and have students get into groups.
2. Provide time for groups to work on the problem and come up with their solutions.
3. Discuss groups' solutions and the methods they used for determining those solutions. Explore one of the extensions, or have students explore extensions of their own.

Connecting Learning
1. What did you have to do in order to solve this problem? [figure out how many of each numeral was needed]
2. What process did your group use to find this number? How does this compare to the processes used by others?
3. Is one easier or more effective than another? Why do you say this?
4. Which store is a better deal? How do you know?
5. What other questions did you think of to explore?

Extensions
1. With the pricing schemes given, find the point at which the stationery store becomes a better deal than the discount store.
2. Determine what the sales tax in your area (if you have any) would add to the total cost of the numerals.

SOLVE IT! 4th © 2012 AIMS Education Foundation

3. Imagine you live in one of the countries in the southern hemisphere that start the school year in February after summer vacation ends. In what month would the hundredth day of school fall? (A related science question is: Why are students in these countries on summer vacation in January?)
4. If you start school in Kindergarten and there are 180 days in each school year, find the grade in which the thousandth day of school would fall.
5. Determine how many days will you spend in school by the time you graduate from high school.

Solutions

In order to solve this problem, students need to break it into several steps. The first step is to determine the total number of cut-out numerals that need to be purchased to make all the numbers from one to 100. This can be done in several different ways. One way is to simply list all the numbers and then count the total number of digits that have been written. A more organized approach might be to go through and figure out how many of each numeral will be needed by counting them one at a time (i.e., 1, 10, 11, 12, 13, ... 21, 31, 41, 51, ... 100). Another approach might be to divide the numbers from one to 100 by place value and find the total number of numerals needed for the ones place and the total needed for the tens place. In the ones place, the numerals zero to nine will each be used 10 times. In the tens place the numeral zero will only be used once (in 100) while the numerals one to nine will each be used 10 times. There will be an additional numeral, one, needed for the hundreds place for a total of 192 (11 zeros, 21 ones, and 20 each of the other numerals from two to nine). Yet another way to calculate the total is to realize that there are 100 numerals in the ones place since every number from one to 100 has a numeral in the ones place; there are 91 numerals in the tens place since there are 10 ones in the 10s, 10 twos in the 20s, 10 threes in the 30s, ... 10 nines in the 90s, and one zero in the 100; and there is one numeral (1) in the hundreds place. Any of these methods is valid and your students may come up with others that have not been listed here. You can make this problem richer for your students by encouraging them to solve this first step in several different ways.

The second step is to determine which store offers a better deal on the cut-out numerals. The cost at the local discount store is found by simple multiplication—192 numerals times five cents is $9.60—well within the students' $20 budget. The cost for the stationery store's numerals is a bit more complex. The first 10 are $0.10 each, or $1.00. The next 100 are $0.05 each, or $5.00. This leaves 82 numerals left to buy at $0.03 each, which totals $2.46. The total cost of $8.46 from the stationery store yields a savings of $1.14 over the discount store.

* © Copyright 2010. National Governors Association Center for Best Practices and Council of Chief State School Officers. All rights reserved.

Centennial Celebration Calculations

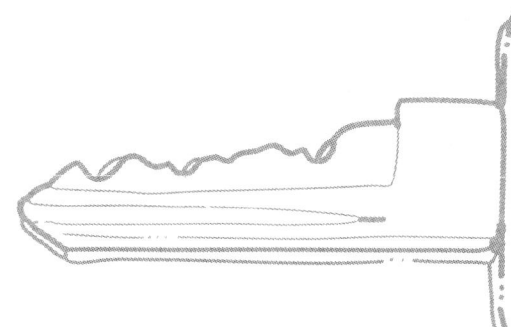

Key Question

Which store provides the best deal for the numerals you need to buy for your celebration?

Learning Goals

Students will:

- determine the number of cut-out numerals that need to purchased to make all of the numbers from one to 100, and

- determine which of two stores offers a better deal for the quantity of numerals needed.

Centennial Celebration Calculations

Your class has been put in charge of decorating the cafeteria for the hundredth day of school celebration. You would like to use fancy cut-out numerals to make all the numbers from one to 100 to place around the tops of the cafeteria walls.

After doing some research you find that the local discount store has gold numerals for 5¢ each. The local stationery store has the exact same numerals, but a different pricing scheme. They charge 10¢ per numeral for the first 10. The next 100 numerals are 5¢ apiece. Any numerals beyond that are 3¢ each.

Your class has a budget of $20 for decorations. Can you buy the numerals as part of your decoration plans? Which store has a better deal? How many of each numeral will you need?

Use the space below to do your work.

Challenge: Try to think of some extensions for this activity. Pick one to do and then report your findings to the class.

Centennial Celebration Calculations

Your class has been put in charge of decorating the cafeteria for the hundredth day of school celebration. You would like to use fancy cut-out numerals to make all the numbers from one to 100 to place around the tops of the cafeteria walls.

After doing some research you find that the local discount store has gold numerals for 5¢ each. The local stationery store has the exact same numerals, but a different pricing scheme. They charge 10¢ per numeral for the first 10. The next 100 numerals are 5¢ apiece. Any numerals beyond that are 3¢ each.

Your class has a budget of $20 for decorations. Can you buy the numerals as part of your decoration plans? Which store has a better deal? How many of each numeral will you need? Use the tables below to help you come up with your answers.

Numeral	Total Number Needed
0	
1	
2	
3	
4	
5	
6	
7	
8	
9	

Discount Store

$0.05 x Total = $ _____

Stationery Store

$0.10 x First 10 = $ _____

$0.05 x Next 100 = $ _____

$0.03 x Any additional = $ _____

Total Cost = $ _____

Why don't you need the same amount of each numeral?

Which numeral do you need the most of? Why?

Which store has the best deal?

How much money will you save at that store?

How much money will that leave you for other decorations?

SOLVE IT! 4th

Centennial Celebration Calculations

Connecting Learning

1. What did you have to do in order to solve this problem?

2. What process did your group use to find this number? How does this compare to the processes used by others?

3. Is one easier or more effective than another? Why do you say this?

4. Which store is a better deal? How do you know?

5. What other questions did you think of to explore?

Problem-Solving Strategies
Guess and Check

Sometimes to solve a problem, it's a good idea to just make a guess. Then you can check your answer to see if it's correct. If it's not, make another guess using what you learned from your first guess. Soon you will find the correct answer. This is a good strategy to use when you don't know how to approach a problem. It's also good when the problem is very complicated or has lots of answers.

Square One

Topic
Logical thinking

Key Question
How can you divide each of the large square grids into smaller squares so that each smaller square contains exactly one star?

Learning Goal
Students will use logical thinking to solve a puzzle that involves dividing a square grid into smaller squares according to certain rules.

Guiding Document
*Common Core State Standards for Mathematics**
- *Make sense of problems and persevere in solving them. (MP1)*

Math
Logical thinking
Problem solving

Integrated Processes
Observing
Comparing and contrasting

Problem-Solving Strategies
Guess and check
Use logical thinking
Use manipulatives

Materials
Student page
Small squares, optional (see *Management 1*)
Scissors, optional

Background Information
This activity is a simplification of a puzzle found in the May 2003 issue of the *Games* magazine. The goal is to divide each of the large square grids into smaller squares so that each smaller square contains exactly one star. The smaller squares will not all be the same size, and no extra spaces are allowed, that is, every part of the large square must be divided into smaller squares.

Management
1. In order for students to solve these problems without going through several erasers, you may want to have them cut the grid paper provided into squares of various sizes. These cut-out squares can then be laid over each large square in different configurations until each square covers exactly one star. Once the solution is discovered, students can draw in the lines dividing the large square, or color each smaller square a different color.
2. Students should work on this problem individually. Depending on how difficult they find it, you may wish to give them small amounts of time to work on it over a period of two or more days.

Procedure
1. Give students a copy of the student page. If desired, provide them with the sheet of grid paper and scissors so that they can cut out squares of different sizes to use while searching for solutions.
2. Allow time for students to find solutions for all of the problems.
3. Have a time of sharing where students compare solutions.
4. Encourage students to share some of the thought processes they had as they solved the problems. Hopefully students will be able to recognize some of the "rules" and logic that guide solving these puzzles (see *Solutions*).

Connecting Learning
1. How did you go about solving the problems?
2. Did you discover any "rules" or helpful places to start when solving? Describe them.
3. Were all of the problems equally difficult? Which ones were harder? Why?

Extension
Have students create their own problems and trade them with classmates to solve. These problems can be created on the grid paper provided. The easiest way to create a problem is to begin by dividing a large square grid into the desired number of smaller squares. Once the large square is sub-divided, fill each smaller square with a single star. Creating the squares before placing the stars assures that a solution will be possible.

SOLVE IT! 4th © 2012 AIMS Education Foundation

Solutions
The solutions for the four puzzles are shown here.

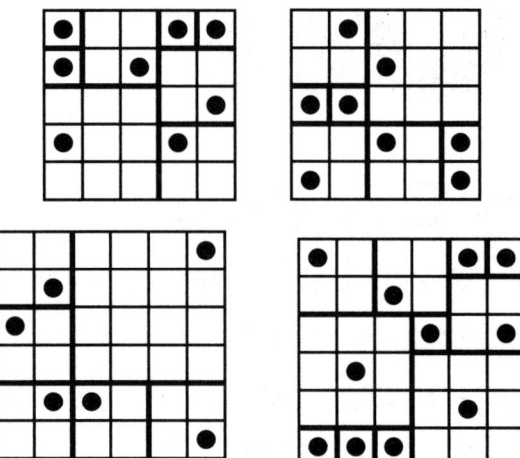

Certain logical "rules" govern the solving of these puzzles. For example, in the first problem, you know that the star in the top left corner of the grid must be in its own 1 x 1 square. You know this because if a 2 x 2 square were placed in that corner, it would contain two stars instead of one.

A two-by-two square
in the corner would
contain two stars

Continuing with this logic, you know that the star just below the one in the top left corner must also be in its own 1 x 1 square because if it were in a 2 x 2 square, it would leave an empty square above it that does not contain a star and cannot be made part of any other square.

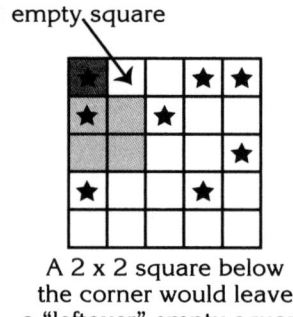

A 2 x 2 square below
the corner would leave
a "leftover" empty square

*   © Copyright 2010. National Governors Association Center for Best Practices and Council of Chief State School Officers. All rights reserved.

Key Question

How can you divide each of the large square grids into smaller squares so that each smaller square contains exactly one star?

Learning Goal

use logical thinking to solve a puzzle that involved dividing a square grid into smaller squares according to certain rules.

Square ★ ne

Your goal is to divide each square below into smaller squares so that every square contains exactly one star. The smaller squares will not all be the same size. The entire square must be divided. There can be no "leftover" spaces. This square shows an example of how one grid would be divided.

Describe how you solved each problem. Were some easier than others? Why or why not?

SOLVE IT! 4th

Connecting Learning

1. How did you go about solving the problems?

2. Did you discover any "rules" or helpful places to start when solving? Describe them.

3. Were all of the problems equally difficult? Which ones were harder? Why?

TIME PIECES

Topic
Problem solving

Key Question
How can you divide the face of a clock into three (and then six) regions so that the sum of the numbers in each region is the same?

Learning Goals
Students will:
- divide a clock face into three regions so that the sum of the numbers in each region is the same, and
- use what they learned to divide a clock face into six regions with the same outcome.

Guiding Document
*Common Core State Standards for Mathematics**
- *Make sense of problems and persevere in solving them. (MP1)*

Math
Problem solving

Integrated Processes
Observing
Comparing and contrasting
Recording

Problem-Solving Strategies
Guess and check
Use manipulatives

Materials
String (see *Management 1*)
Student page

Background Information
This puzzle is an adaptation of one found in *The Moscow Puzzles: 359 Mathematical Recreations* by a Russian high school teacher named Boris A. Kordemsky. The English translation of this book was edited by Martin Gardner and is published by Dover Publications.

There are two challenges in *Time Pieces*. The first is to divide a standard clock face into three regions so that the sum of the numbers in each region is the same. The second is to divide the clock face into six regions with the sums of the numbers in each region being equal.

While this puzzle can be solved by trial and error, a more effective way is to use a mathematical approach. Calculating the sum of all 12 numbers on the clock face will help you approach the puzzle in a more mathematical fashion. This makes finding the target sums for each of the two versions of the puzzle a bit easier. Once this insight is gained, the problem quickly enters a problem-solving arena and has value beyond its attraction as an interesting puzzle.

Management
1. Making pieces of string available for students to use when dividing the clock face into regions may help them solve it. This way students are not repeatedly drawing and erasing lines until they find the solutions. After students find the solutions with the string, they can make a permanent record by drawing in the lines.
2. Students should work on this puzzle individually, and then share their solutions and thought processes with the class.

Procedure
1. Distribute the student page and pieces of string to those who want it.
2. Provide time for students to work on the puzzle and come up with their solutions.
3. Have students share their solutions and the methods they used to come up with these solutions.

Connecting Learning
1. How can you divide the clock into three sections so that the sum of the numbers in each section is equal?
2. How did you find this solution? How does this compare to the ways your classmates found their solutions?
3. How can you divide the clock into six sections so that the sum of the numbers in each section is equal?
4. What other number of sections could you divide a clock face into so that the numbers in all of the sections have the same sum? How do you know?

Solutions

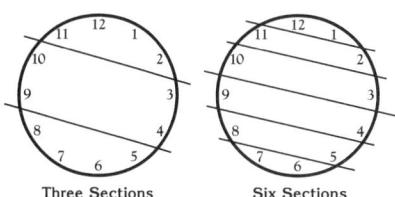

Three Sections   Six Sections

* © Copyright 2010. National Governors Association Center for Best Practices and Council of Chief State School Officers. All rights reserved.

TIME PIECES

Key Question

How can you divide the face of a clock into three (and then six) regions so that the sum of the numbers in each region is the same?

Learning Goals

Students will:

- divide a clock face into three regions so that the sum of the numbers in each region is the same, and

- use what they learned to divide a clock face into six regions with the same outcome.

SOLVE IT! 4th   © 2012 AIMS Education Foundation

TIME PIECES

The first challenge in this puzzle is to draw in lines on the clock face below so that the face is divided into three regions in such a way that the sum of the numbers in each region is the same.

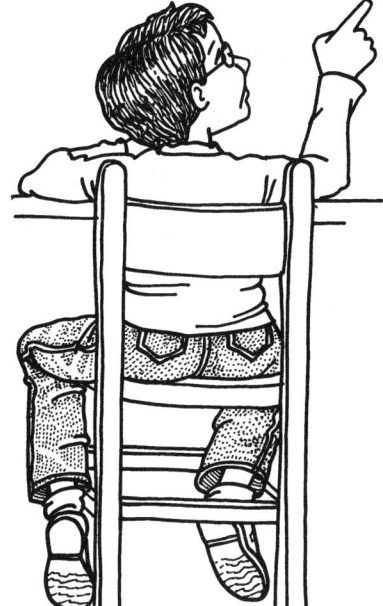

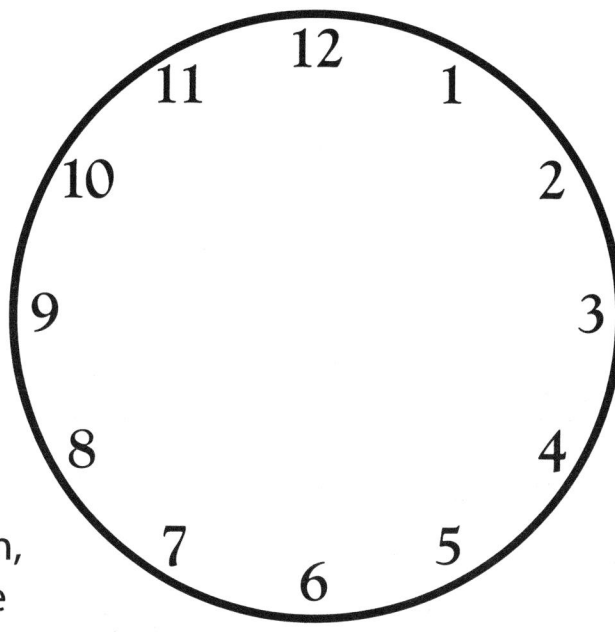

The second challenge is similar to the first. This time the lines you draw must divide the clock face into six regions. Again, the numbers in each region must have the same sum.

Reflect on your solutions. What mathematical discoveries did you make?

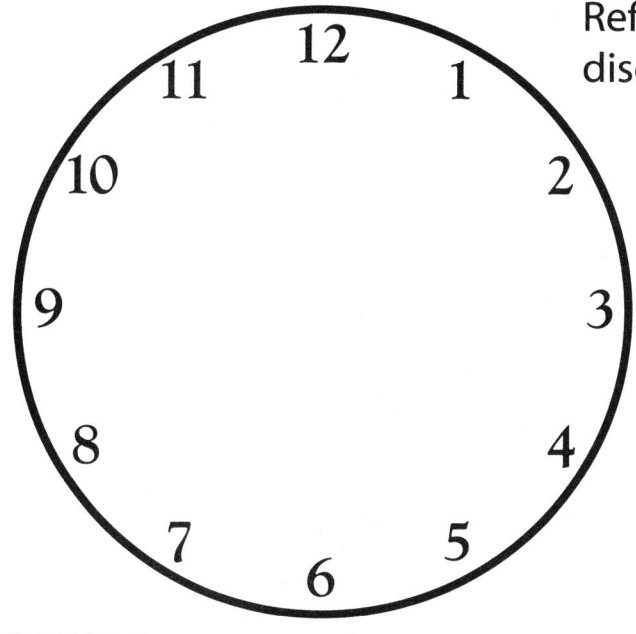

TIME PIECES

Connecting Learning

1. How can you divide the clock into three sections so that the sum of the numbers in each section is equal?

2. How did you find this solution? How does this compare to the ways your classmates found their solutions?

3. How can you divide the clock into six sections so that the sum of the numbers in each section is equal?

4. What other number of sections could you divide a clock face into so that the numbers in all of the sections have the same sum? How do you know?

X-cellent Addition

Topic
Mathematical microworlds

Key Question
How can you place the numbers from one to eight in a double-X arrangement so that the sum of the numbers in each of the four diagonals is the same?

Learning Goals
Students will:
- explore a mathematical microworld,
- find and record all possible solutions for this microworld, and
- look for patterns in their solutions.

Guiding Documents
Project 2061 Benchmark
- *Mathematics is the study of many kinds of patterns, including numbers and shapes and operations on them. Sometimes patterns are studied because they help to explain how the world works or how to solve practical problems, sometimes because they are interesting in themselves.*

Common Core State Standards for Mathematics*
- *Make sense of problems and persevere in solving them. (MP1)*
- *Look for and make use of structure. (MP7)*
- *Generate and analyze patterns. (4.OA)*

Math
Number and operations
   addition
Patterns and relationships
Mathematical microworlds
Problem solving

Integrated Processes
Observing
Comparing and contrasting
Recording
Organizing
Generalizing

Problem-Solving Strategies
Guess and check
Use manipulatives
Look for patterns

Materials
Number cards (see *Management 1*)
Scissors
Student pages

Background Information
This activity is a mathematical microworld. A microworld, in the context of mathematics, is a mathematical environment governed by simple rules and structures. Because the microworld presented here is one that has not been explored before, your students have the opportunity to make exciting and completely new discoveries as they obtain solutions and search for patterns.

X-cellent Addition challenges students to place the numbers from one to eight in a double-X arrangement so that the sum of the numbers on each of the four diagonals is the same. They are then challenged to explore the patterns in the problem and extend it by asking their own related questions.

Management
1. Each student will need eight number cards that will fit in the spaces on the student page. A sheet of 8.5" x 11" paper folded in eighths horizontally and vertically will give eight sets of cards in the right size.
2. This activity is divided into two parts. Two different approaches to the second section have been presented: an open-ended approach, and a structured one. In the open-ended approach (student page three), students are simply asked to write down all of the interesting things that they discover about their solutions. In the structured approach (student pages four and five), two pages of questions are given to guide students' discovery of some of the more interesting and obvious patterns. Choose the approach that is best for your class and hand out the appropriate pages when students are ready for them.
3. It is recommended that you spread this activity over several days, providing small amounts of time for students to work on it.
4. Students will need two copies of the *Solutions* page in order to record all of the possible solutions.

Procedure
1. Distribute paper for the number cards and have students cut out sets for themselves and number them from one to eight.
2. Give each student a copy of the first student page and two *Solutions* pages. You may have students work individually or in small groups of three or four.

SOLVE IT! 4th   © 2012 AIMS Education Foundation

3. Allow time for students to discover and record all of the solutions they can. They should discover at least half of the solutions (six) before moving on to the final student page(s).
4. Distribute the appropriate student page(s) for *Part Two* and provide time for students to respond (see *Management 2*).
5. Once students have analyzed their solutions, come together for a time of class discussion.
6. Have students/groups share the solutions they discovered and record these on the board or on a transparency of the *Solutions* page. Try to identify and eliminate those solutions that are merely flips and/or rotations of each other.
7. Discuss the patterns that students discovered and explore any that they may have missed as a class.
8. If desired, have students extend the problem by exploring related questions of their choosing.

Connecting Learning
1. How many solutions were you able to discover? Do you think you have found them all? Why or why not?
2. What numbers were you able to get as sums?
3. Do you think you got all of the possible sums using the numbers one to eight? Why or why not?
4. What do you notice about the two numbers that are in the middle of your different solutions?
5. What do you notice about the numbers that are in the top row and the bottom row of your different solutions?
6. What patterns do you see if you look at the sums of the middle four numbers and the outer four numbers?
7. What patterns do you notice in the sums and differences of the numbers in the three vertical columns?
8. What other patterns can you find in your solutions?
9. What related questions can you think of to explore?

Extensions
1. Use the numbers two through nine, consecutive odd numbers, or consecutive even numbers.
2. Extend the shape to a triple X and use the numbers one to 11.

Solutions
Students will discover (with persistence) that there are four possible sums that can be achieved with the numbers one to eight—12, 13, 14, and 15. The sums of 13 and 14 both have four unique solutions, but the sums of 12 and 15 each have two unique solutions. (A solution is unique if it is different from the others when it is flipped and/or rotated.)

[Diagrams of solutions with Sum: 12, 13, 14, 15]

The various parts of the solutions that will be examined are defined here using one solution as an example. Pay special attention to the difference between "center numbers" and "inside numbers."

[Diagram: 1, 4, 6 / 8, 3 / 2, 5, 7]
Sum: 14
Vertical columns: 1 & 2, 4 & 5, 6 & 7
Center numbers: 8, 3
Corner numbers: 1, 6, 2, 7
Inside numbers: 4, 8, 3, 5
Top row: 1, 4, 6
Bottom row: 2, 5, 7

Patterns in Differences
- Within any valid solution, the difference between the numbers in each vertical column will always be the same. In the list of solutions given, the only possible differences are one, three and five. For example, those solutions in the first row all have a difference of three between the numbers in the vertical columns (7 – 4 = 3, 6 – 3 = 3, 8 – 5 = 3, 5 – 2 = 3, 4 – 1 = 3), those in the second row all have a difference of five (7 – 2 = 5, 6 – 1 = 5, 8 – 3 = 5), and those in the bottom two rows all have a difference of one (8 – 7 = 1, 3 – 2 = 1, 5 – 4 = 1, etc.)
- Additionally, the difference between the two center numbers must also be either one, three or five. In the first two rows, the difference between the two center numbers is one (2 – 1 = 1, 8 – 7 = 1, 5 – 4 = 1), in the third row the difference is three (6 – 3 = 3, 8 – 5 = 3, 4 – 1 = 3), and in the fourth row it is five (6 – 1 = 5, 8 – 3 = 5).
- When the vertical and center differences are looked at in combination, four possibilities emerge for valid solutions:
  1. Center difference—one; Vertical difference—three; Sums—12, 13, 14, 15
  2. Center difference—one; Vertical difference—five; Sums—13, 14

3. Center difference—three; Vertical difference—one; Sums—12, 13, 14, 15
4. Center difference—five; Vertical difference—one; Sums—13, 14

Patterns in Sums

- For solutions that have a difference of one between their center numbers, the sums of the vertical columns are always odd numbers between five and 13. Within one solution, the sums of the columns are always three consecutive odd numbers, although they are not necessarily in consecutive order as they appear in the solution.
- For solutions with a difference other than one between their center numbers, the sums of the vertical columns are still odd, but they are not *consecutive* odd numbers.
- In order for the sum of every vertical column to be odd, there must always be an odd number and an even number paired together (odd + odd = even, even + even = even, odd + even = odd). It follows then that the top (or bottom) row must always have two odd numbers and an even number, and the bottom (or top) row must always have two even numbers and an odd number. If this were not the case it would be impossible to have both an odd and an even number in the same vertical column. This also means that the two center numbers will never be two evens or two odds, but always one of each. This explains why the difference between the two center numbers can only be one, three or five.
- When the sum of the diagonals is 12, the four corner numbers have a sum of 24, and the four inside numbers have a sum of 12.
- When the sum of the diagonals is 15, the four corner numbers have a sum of 12, and the four inside numbers have a sum of 24.
- When the sum of the diagonals is 13, the four corner numbers have a sum of 20, and the four inside numbers have a sum of 16.
- When the sum of the diagonals is 14, the four corner numbers have a sum of 16, and the four inside numbers have a sum of 20.

Relationships Between Solutions

Take any solution and substitute the letters A to H for the numbers. Any valid solution can be made into another valid solution by rearranging the letters as follows:
1. Replace the center vertical letters (B and G) with the center horizontal letters (D and E) and vice versa.
2. Switch the top left and bottom right letters (A and H).

This leaves you with the arrangement shown here, which will always be a valid solution provided the first solution was valid.

For the purposes of discussion, we will call the original solution (A, B, C, etc.) and the second solution (H, D, C, etc.) partner solutions. There are a total of six sets of partner solutions. Both solutions within a partner set have the same sum. When you examine the differences between the center numbers and differences within the vertical columns in the two partner solutions, you will find that they are exact opposites. For example, the two solutions with a sum of 12 have center differences of three and one, and vertical differences of one and three respectively.

Sum: 12
Center differences: 1, 3
Vertical differences: 3, 1

Sum: 12

Sum: 13
Center differences: 1, 3
Vertical differences: 3, 1

Sum: 13

Sum: 13
Center differences: 1, 5
Vertical differences: 5, 1

Sum: 13

Sum: 14
Center differences: 1, 5
Vertical differences: 5, 1

Sum: 14

Sum: 14
Center differences: 1, 3
Vertical differences: 3, 1

Sum: 14

Sum: 15
Center differences: 1, 3
Vertical differences: 3, 1

Sum: 15

* © Copyright 2010. National Governors Association Center for Best Practices and Council of Chief State School Officers. All rights reserved.

SOLVE IT! 4th   153   © 2012 AIMS Education Foundation

X-cellent Addition

Key Question

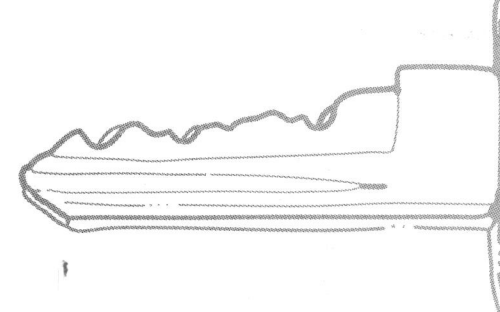

How can you place the numbers from one to eight in a double-X arrangement so that the sum of the numbers in each of the four diagonals is the same?

Learning Goals

Students will:

- explore a mathematical microworld,

- find and record all possible solutions for this microworld, and

- look for patterns in their solutions.

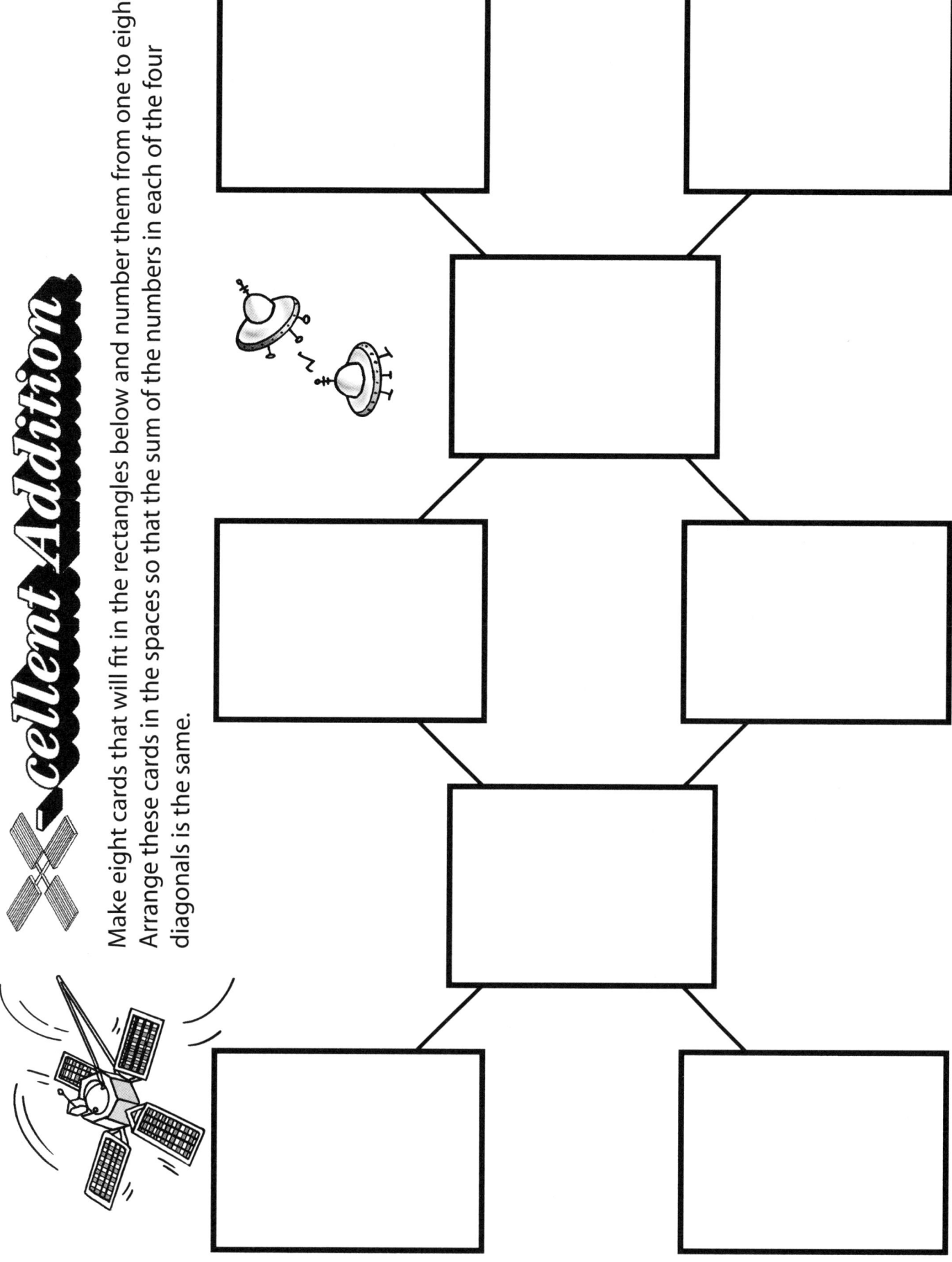

X-cellent Addition

Make eight cards that will fit in the rectangles below and number them from one to eight. Arrange these cards in the spaces so that the sum of the numbers in each of the four diagonals is the same.

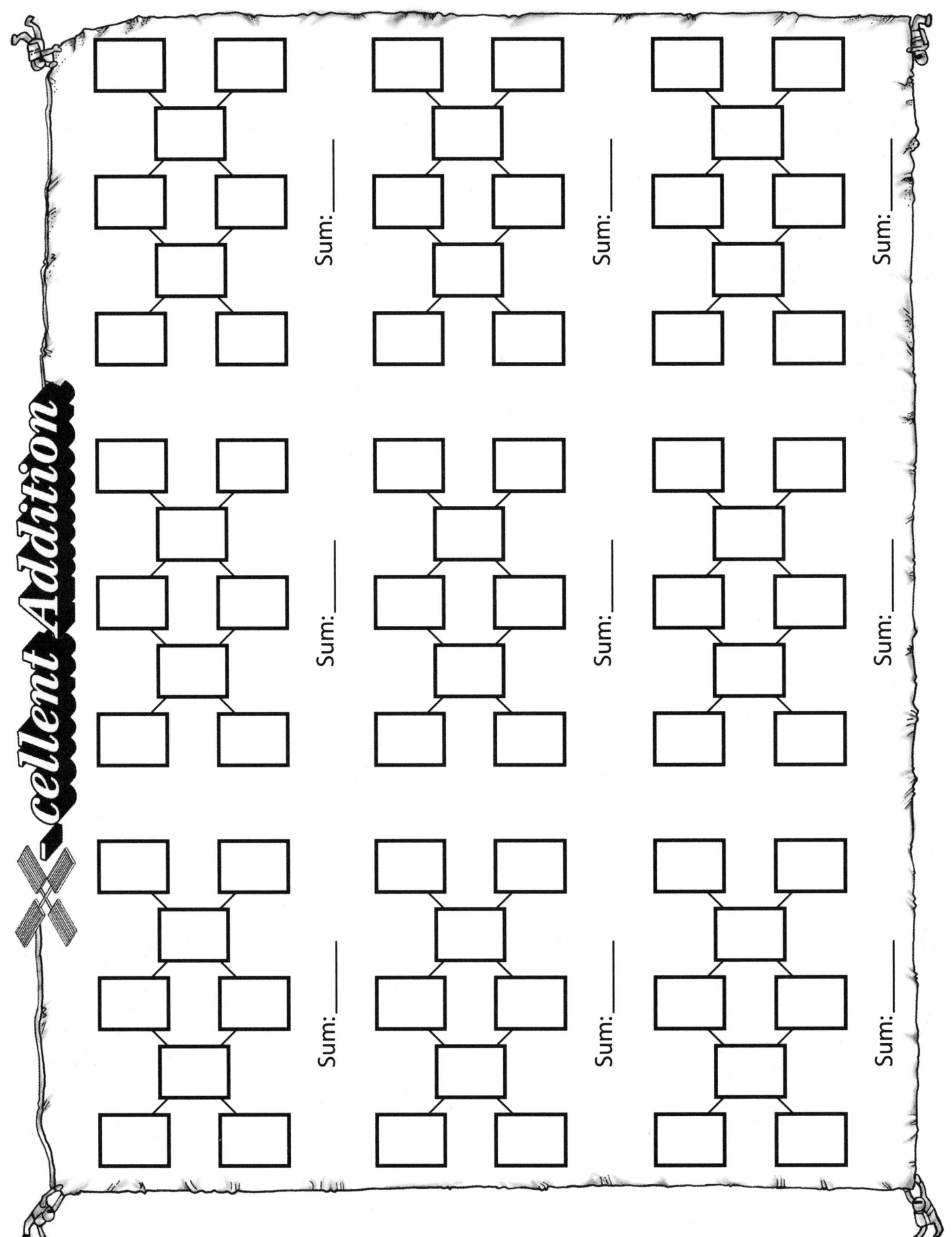

X-cellent Addition

Study your solutions carefully. How many are actually different? What sums were you able to get? What patterns do you notice? Write down all of the interesting things you can discover about your solutions.

Think of a question related to this problem that you would like to explore. Write it below, explore it, and record your findings.

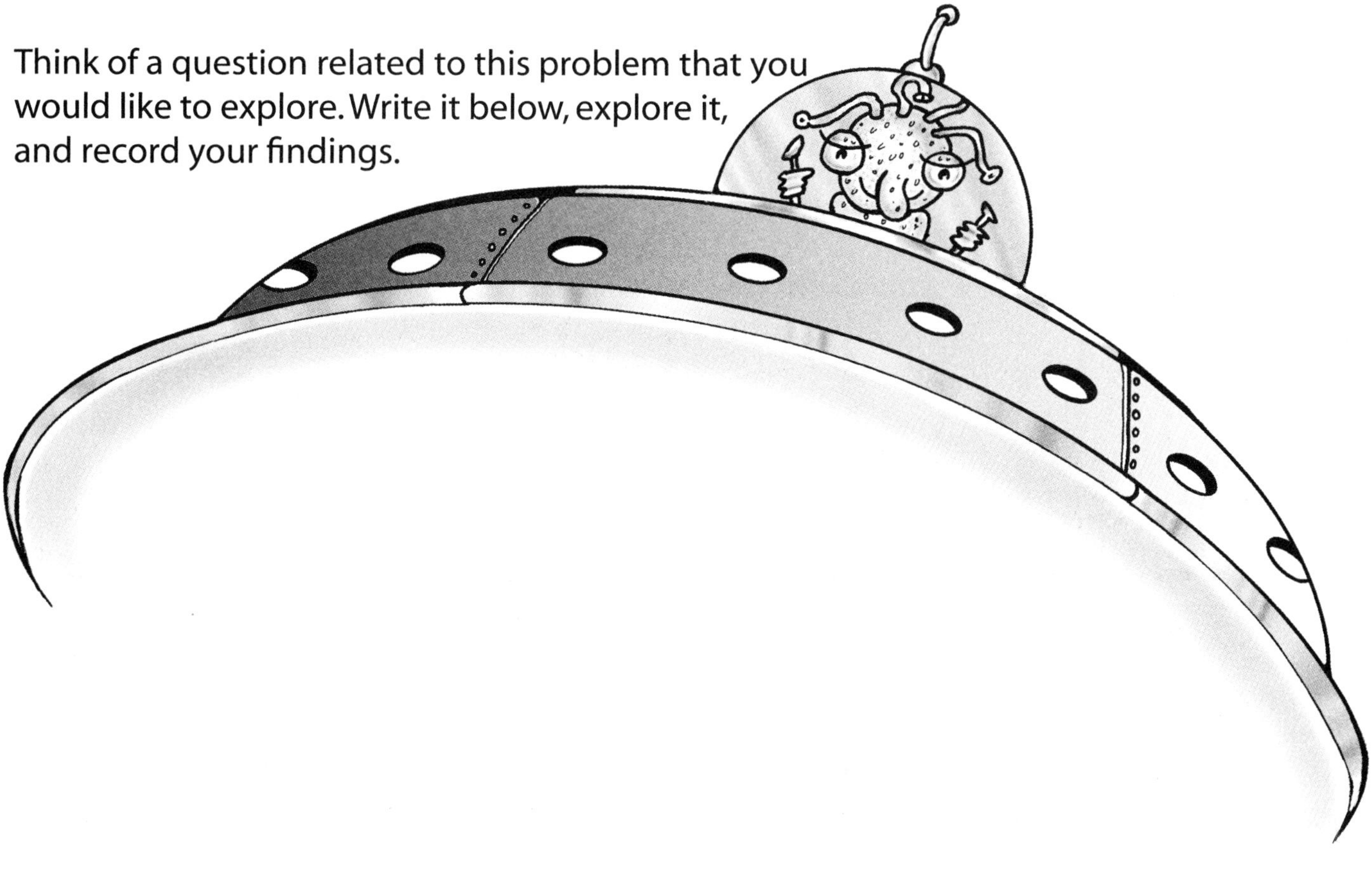

X-cellent Addition

When you have found at least six different solutions, answer the following questions.

1. Look at your solutions carefully. If you don't count rotations and flips of the same arrangement, how many unique solutions were you able to come up with?

2. What numbers were you able to get as sums?

3. Do you think you got all of the possible sums using the numbers one to eight? Why or why not?

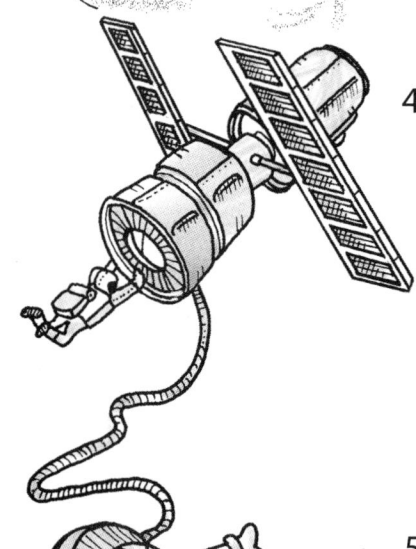

4. What do you notice about the two numbers that are in the middle of your different solutions?

5. What do you notice about the numbers that are in the top row and the bottom row of your different solutions?

SOLVE IT! 4th © 2012 AIMS Education Foundation

X-cellent Addition

6. What patterns do you see if you look at the sums of the middle four numbers and the outer four numbers?

7. What patterns do you notice in the sums and differences of the numbers in the three vertical columns?

8. What other patterns can you find in your solutions?

9. Write down several questions related to this problem that you think would be interesting to explore.

10. Pick one of your questions and explore it. Report your findings.

X-cellent Addition

Connecting Learning

1. How many solutions were you able to discover? Do you think you have found them all? Why or why not?

2. What numbers were you able to get as sums?

3. Do you think you got all of the possible sums using the numbers one to eight? Why or why not?

4. What do you notice about the two numbers that are in the middle of your different solutions?

5. What do you notice about the numbers that are in the top row and the bottom row of your different solutions?

X-cellent Addition

Connecting Learning

6. What patterns do you see if you look at the sums of the middle four numbers and the outer four numbers?

7. What patterns do you notice in the sums and differences of the numbers in the three vertical columns?

8. What other patterns can you find in your solutions?

9. What related questions can you think of to explore?

Problem-Solving Strategies
Look for Patterns

Patterns are everywhere. Some patterns repeat. Other patterns grow. Some patterns use numbers. Some patterns use shapes. Some patterns are cycles. Looking for patterns can help you solve problems. Sometimes, knowing the pattern and how to extend it gives you the answer.

The Fabulous Four-Sum

Topic
Problem solving

Key Question
Using the numbers one to four and two to five, how many different addition problems can you create when the numbers are arranged in a two-by-two array?

Learning Goals
Students will:
* determine different possible addition problems using a specified set of numbers,
* look for patterns within their solutions, and
* look for patterns between solutions that use different sets of numbers.

Guiding Documents
Project 2061 Benchmark
* *Mathematics is the study of many kinds of patterns, including numbers and shapes and operations on them. Sometimes patterns are studied because they help to explain how the world works or how to solve practical problems, sometimes because they are interesting in themselves.*

Common Core State Standards for Mathematics*
* *Make sense of problems and persevere in solving them. (MP1)*
* *Look for and make use of structure. (MP7)*

Math
Computation
　addition
Pattern recognition
Problem solving

Integrated Processes
Observing
Comparing and contrasting
Collecting and recording data
Organizing data
Generalizing

Problem-Solving Strategies
Look for patterns
Organize the information
Use manipulatives

Materials
Number cards (see *Management 1*)
Student pages
Small sticky notes, optional (see *Management 3*)

Background Information
　　This activity is one that combines basic computation with problem solving and pattern recognition. It is a rich problem with many opportunities for extensions and in-depth study, and can be explored at different levels of sophistication, depending on the abilities of your students.
　　The initial problem is a simple one: using the numbers one through four and two through five, how many different addition problems can you create when the numbers are arranged in a two-by-two array? Once the solutions for each set of numbers are discovered, the challenge is to search for patterns within and between the sets of solutions.

Management
1. In order to help students discover the many possibilities, they should each make a set of number cards (1–5) from scratch paper that will fit into the spaces provided on the first student sheet.
2. There are five student pages for this activity. The first page explains the problem and provides spaces in which students can manipulate their number cards. The second and third student pages provide spaces for recording solutions, and the fourth and fifth pages have exploration questions to guide students' pattern discoveries. The final pages should not be distributed to students until they believe they have discovered all of the solutions for both sets of numbers. Depending on the ages and abilities of your students, this may take several days.
3. Students should be encouraged to organize their solutions in a logical fashion. In order to facilitate this process, it is recommended that you provide students with small sticky notes on which to write their solutions prior to recording them on the solutions pages. This will allow them to experiment with a variety of organizational schemes without having to erase and re-write their answers.

SOLVE IT! 4ᵗʰ　　　　© 2012 AIMS Education Foundation

Procedure
1. Distribute scratch paper, and have students make number cards from one to five.
2. Give each student a copy of the first student page and the recording pages. Go over the instructions as a class and be sure that everyone understands the challenges.
3. If desired, provide students with small sticky notes on which to record their solutions before writing them on their recording pages.
4. Inform students that for the purposes of this activity, solutions should be considered unique even if they are the same numbers in a different order. For example, 24 + 13 and 13 + 24 would each be a unique solution.
5. Once students have had ample time to discover, organize, and record solutions for both challenges, distribute the final two student pages.
6. Close with a time of class discussion where students can share their solutions and explore the patterns they discovered.

Connecting Learning
1. What sums did you discover using the numbers one to four? Do you think you have found them all? Why or why not?
2. What sums did you discover using the numbers two to five? Do you think you have found them all? Why or why not?
3. Look at the sums using the numbers from one to four and from two to five. What patterns do you see in the sets of numbers?
4. How can you explain these patterns?
5. How many times does each sum occur? Why? How would this number be different if the same two numbers in a different order (for example, 14 + 23 and 23 + 14) were considered to be the same solution?
6. What patterns do you notice in the numbers that add up to the same sum using the digits one to four?
7. What patterns do you notice in the numbers that add up to the same sum using the digits two to five?
8. How do the sums using the digits one to four compare to the sums using the digits two to five?
9. Using the patterns you observed, predict what the sums would be if you used the numbers three through six.
10. Describe any other patterns you discovered while exploring this problem.

Extensions
1. Use the numbers from three to six and four to seven and compare the patterns to those discovered in this activity. (Some of the patterns do not hold for the larger numbers, but other exciting patterns begin to develop in their place.)
2. Find all of the possible products using the numbers one to four.
3. Find all of the possible differences (positive and negative) using the numbers one to four and two to five.
4. Use the numbers from one to five in a three by two array and determine all of the possible sums.

Solutions
Using the rules given, there are 24 unique solutions for each set of numbers. These solutions are shown here using two different organizational schemes.

Organized by the Value of the Top Addend
Using the numbers 1, 2, 3, and 4

12	12	13	13	14	14
+34	+43	+24	+42	+23	+32
46	55	37	55	37	46
21	21	23	23	24	24
+34	+43	+14	+41	+13	+31
55	64	37	64	37	55
31	31	32	32	34	34
+24	+42	+14	+41	+12	+21
55	73	46	73	46	55
41	41	42	42	43	43
+23	+32	+13	+31	+12	+21
64	73	55	73	55	64

Organized by the Sum
Using the numbers 1, 2, 3, and 4

13	14	23	24	12	14	32	34
+24	+23	+14	+13	+34	+32	+14	+12
37	37	37	37	46	46	46	46
12	13	21	24	31	34	42	43
+43	+42	+34	+31	+24	+21	+13	+12
55	55	55	55	55	55	55	55
21	23	41	43	31	32	41	42
+43	+41	+23	+21	+42	+41	+32	+31
64	64	64	64	73	73	73	73

Organized by the Value of the Top Addend
Using the numbers 2, 3, 4, and 5

23	23	24	24	25	25	32	32
+45	+54	+35	+53	+34	+43	+45	+54
68	77	59	77	59	68	77	86
34	34	35	35	42	42	43	43
+25	+52	+24	+42	+35	+53	+25	+52
59	86	59	77	77	95	68	95
45	45	52	52	53	53	54	54
+23	+32	+34	+43	+24	+42	+23	+32
68	77	86	95	77	95	77	86

SOLVE IT! 4th

Organized by the Sum
Using the numbers 2, 3, 4, and 5

```
  24     25     34     35     23     25     43     45
+ 35   + 34   + 25   + 24   + 45   + 43   + 25   + 23
  59     59     59     59     68     68     68     68

  23     24     32     35     42     45     53     54
+ 54   + 53   + 45   + 42   + 35   + 32   + 24   + 23
  77     77     77     77     77     77     77     77

  32     34     52     54     42     43     52     53
+ 54   + 52   + 34   + 32   + 53   + 52   + 43   + 42
  86     86     86     86     95     95     95     95
```

What follows is a discussion of some of the patterns suggested by the questions on the final student pages, as well as other patterns that your students may discover.

- There are five possible sums for each set of numbers. The sums for the numbers one to four are 37, 46, 55, 64, and 73. The sums for the numbers two to five are 59, 68, 77, 86, and 95.
- It is simple to determine if all sums have been discovered because there is no carrying involved in the addition. If every two-number combination that can be created using the digits is summed, the possible values of the digits in the solutions are known.

  Numbers 1-4: $1 + 2 = 3$ $1 + 3 = 4$ $1 + 4 = 5$
  $2 + 3 = 5$ $2 + 4 = 6$ $3 + 4 = 7$

  Numbers 2-5: $2 + 3 = 5$ $2 + 4 = 6$ $2 + 5 = 7$
  $3 + 4 = 7$ $3 + 5 = 8$ $4 + 5 = 9$

  Once these values are determined, they can be placed together based on the numbers they use. For example, using the numbers two through five, the sum of 2 and 3 must go with the sum of 4 and 5 because each digit can only be used once. This pairing produces the sums of 59 and 95.

- The sum of the digits in each solution using the numbers one to four is 10. ($3 + 7 = 10$, $4 + 6 = 10$, $5 + 5 = 10$) The sum of the digits in each solution using the numbers two to five is 14. ($5 + 9 = 14$, $6 + 8 = 14$, $7 + 7 = 14$)

- When arranged from least to greatest, the sums increase by nine. ($37 + 9 = 46$, $46 + 9 = 55$, etc. $59 + 9 = 68$, $68 + 9 = 77$, etc.)

- For the digits one to four, the sums 37, 46, 64, and 73 occur four times each, and the sum 55 occurs eight times. For the digits two to five, the sums 59, 68, 86, and 95 occur four times each, and the sum 77 occurs eight times. These values would be reduced by half if the same two numbers in a different order were considered to be the same solution.

- Each sum has a "partner" that contains the same digits in the opposite order. The sums 37 and 73 are partners, 46 and 64 are partners, 59 and 95 are partners, and 68 and 86 are partners. As mentioned, each of these sums occurs four times, meaning that each set of partner sums occurs eight times. The sums 55 and 77 are both palindromes—the same read forward or backward—and are partners with themselves, which is the reason they each occur eight times.

- Because there is no carrying involved in the addition, the numbers that add up to the same sum are very straightforward. Using the numbers one to four for example, the sum 37 is produced by numbers in the 10s and the 20s (13 or 14 plus 23 or 24) and the sum 46 is produced by numbers in the 10s and 30s (12 or 13 plus 32 or 34). Likewise, using the numbers two through five, the sum 59 is produced by numbers in the 20s and 30s (24 + 35 or 34 + 25) and the sum 86 is produced by numbers in the 30s and 50s (32 + 54 or 34 + 52).

- The sums produced using the numbers two through five are 22 greater than the sums produced using the numbers one through four.
  $37 + 22 = 59$ $46 + 22 = 68$
  $55 + 22 = 77$ $64 + 22 = 86$
  $73 + 22 = 95$

* © Copyright 2010. National Governors Association Center for Best Practices and Council of Chief State School Officers. All rights reserved.

Key Question

Using the numbers one to four and two to five, how many different addition problems can you create when the numbers are arranged in a two by two array?

Learning Goals

- determine different possible addition problems using a specified set of numbers,

- look for patterns within their solutions, and

- look for patterns between solutions that use different sets of numbers.

The Fabulous Four-Sum

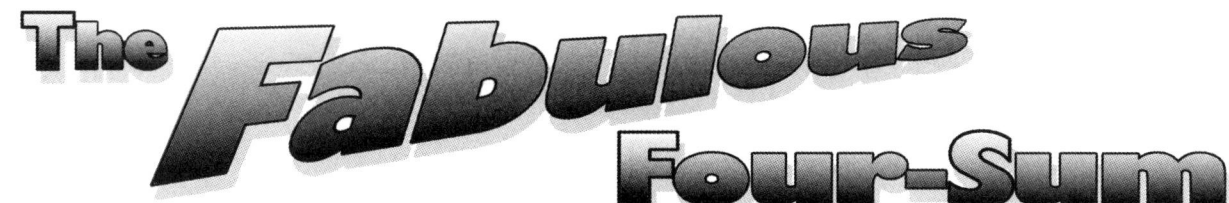

Part One

How many different ways can you make an addition problem by arranging the numbers 1, 2, 3, and 4 in the spaces below? Use your number cards to discover as many different combinations as you can and record them (and their sums) in the spaces provided on the first solutions page. Try to develop a systematic approach and organize your solutions in a logical fashion. The same two numbers in a different order should be recorded as two different solutions. For example, 12 + 34 = 46 and 34 + 12 = 46 are both unique solutions.

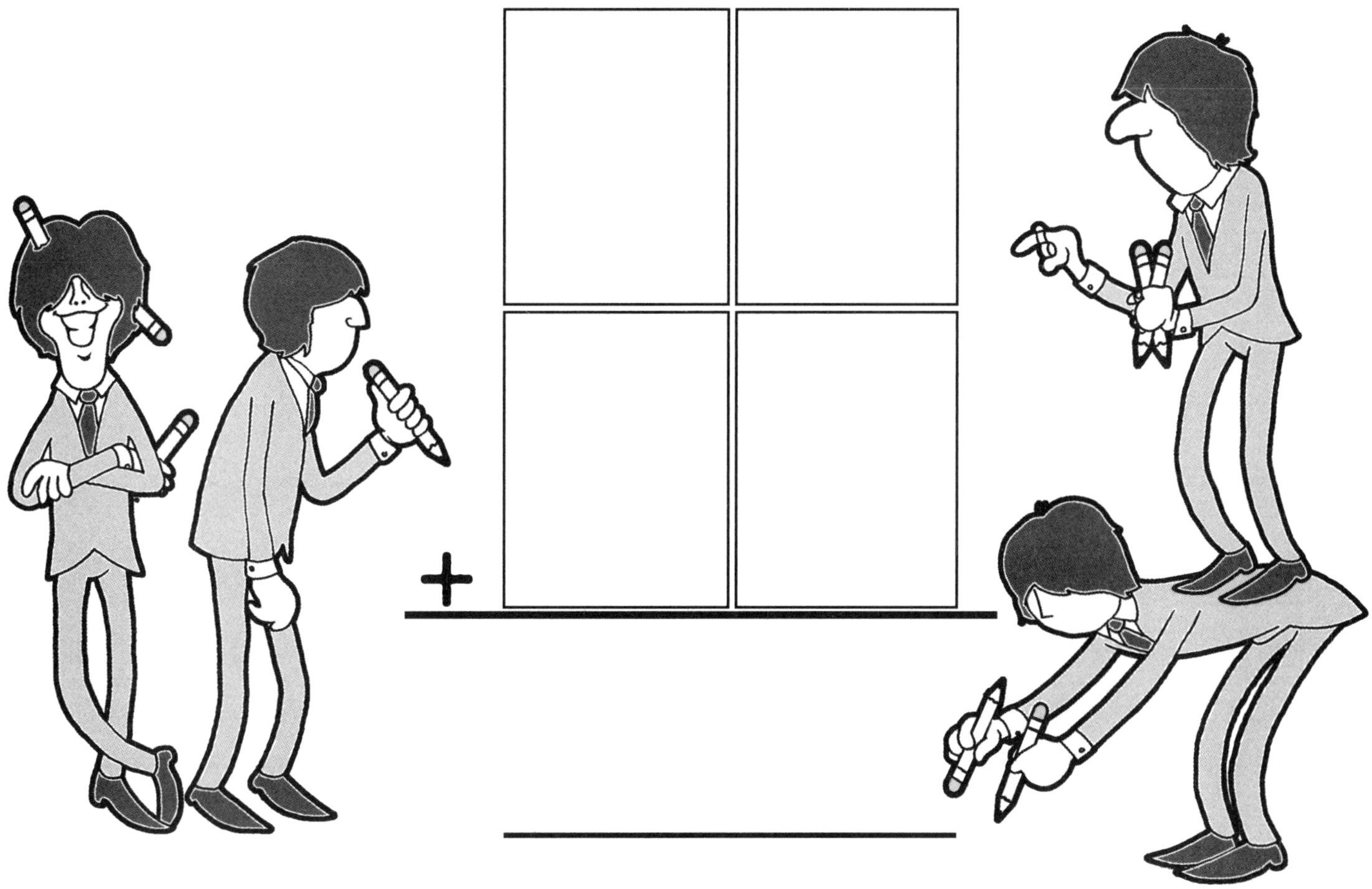

Part Two

Follow the same procedure described above, but use the numbers 2, 3, 4, and 5. Record all of the solutions you can find in the spaces on the second solutions page.

Solutions—Part One

Record each solution you discover using the numbers one through four below, as well as the sum produced by the numbers. Try to develop a systematic approach and organize your solutions in a logical order.

SOLVE IT! 4th   170   © 2012 AIMS Education Foundation

The Fabulous Four-Sum

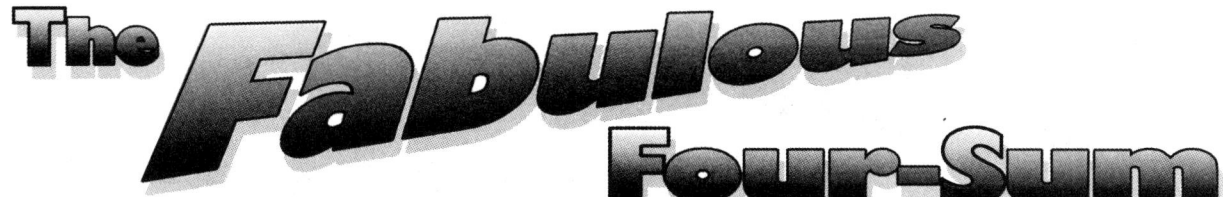

Solutions—Part Two

Record each solution you discover using the numbers two through five below, as well as the sum produced by the numbers. Try to develop a systematic approach and organize your solutions in a logical order.

The Fabulous Four-Sum

Use the questions below to help you think about your solutions and discover some of the patterns they contain.

1. What sums did you discover using the numbers one to four? Do you think you have found them all? Why or why not?

2. What sums did you discover using the numbers two to five? Do you think you have found them all? Why or why not?

3. Look at the sums in the previous two answers. What patterns do you see in the numbers?

4. What are the reasons for the patterns in the sums?

5. How many times does each sum occur? Why? How would this number be different if the same two numbers in a different order (for example, 14 + 23 and 23 + 14) were considered to be the same solution?

SOLVE IT! 4th © 2012 AIMS Education Foundation

The Fabulous Four-Sum

6. What patterns do you notice in the numbers that add up to the same sum using the digits one to four?

7. What patterns do you notice in the numbers that add up to the same sum using the digits two to five?

8. How do the sums using the digits one to four compare to the sums using the digits two to five?

9. Using the patterns you observed, predict what the sums would be if you used the numbers three through six.

10. Describe any other patterns you discovered while exploring this problem.

Connecting Learning

1. What sums did you discover using the numbers one to four? Do you think you have found them all? Why or why not?

2. What sums did you discover using the numbers two to five? Do you think you have found them all? Why or why not?

3. Look at the sums using the numbers from one to four and from two to five. What patterns do you see in the sets of numbers?

4. How can you explain these patterns?

5. How many times does each sum occur? Why? How would this number be different if the same two numbers in a different order (for example, 14 + 23 and 23 + 14) were considered to be the same solution?

The Fabulous Four-Sum

Connecting Learning

6. What patterns do you notice in the numbers that add up to the same sum using the digits one to four?

7. What patterns do you notice in the numbers that add up to the same sum using the digits two to five?

8. How do the sums using the digits one to four compare to the sums using the digits two to five?

9. Using the patterns you observed, predict what the sums would be if you used the numbers three through six.

10. Describe any other patterns you discovered while exploring this problem.

Crack the Code

Topic
Pattern recognition

Key Question
What are the missing numbers in the tables of codes?

Learning Goal
Students will identify the missing numbers in tables of codes.

Guiding Documents
Project 2061 Benchmark
- Mathematics is the study of many kinds of patterns, including numbers and shapes and operations on them. Sometimes patterns are studied because they help to explain how the world works or how to solve practical problems, sometimes because they are interesting in themselves.

Common Core State Standards for Mathematics*
- Make sense of problems and persevere in solving them. (MP1)
- Look for and make use of structure. (MP7)
- Generate and analyze patterns. (4.OA)

Math
Patterns
Problem solving

Integrated Processes
Observing
Comparing and contrasting
Recording

Problem-Solving Strategy
Look for patterns

Materials
Student page

Background Information
This activity is designed to be done in tandem with the one on the following pages. To get the maximum benefit from this experience, it is recommended that you follow up *Crack the Code* with *Code Consultant*. These activities will challenge students to exercise their problem-solving and pattern-recognition skills in an interesting setting.

The premise for *Crack the Code* is a new "smart alarm" that has been created by the (fictitious) company *Burgle-Tech*. Smart alarms have codes that change every day based on a pattern. Some of the codes are two-digit, some are three-digit, and some are four-digit. The challenge for students is to discover the patterns and use them to fill in the missing data. This kind of "fill-in-the-blank" problem solving gives students practice with identifying and generalizing patterns and is great experience for standardized tests.

Management
1. Students should work individually to try and discover the codes.
2. This activity should be done before *Code Consultant*.

Procedure
1. Distribute the student page and go over the challenge as a class.
2. Provide time for students to work individually on the problems.
3. When students have had time to complete all four tables, come together for a time of class discussion.
4. Have students share their answers and how they determined them. Where there are discrepancies, determine the correct response.

Connecting Learning
1. What are the missing codes? (See *Solutions*.)
2. How were you able to find the missing codes?
3. Which codes were the most difficult to crack? Why do you think this is?
4. What would the code be for the 15th in the first table? [30]
5. What would the code be for today's date in today's year in the last table?
6. Describe the rule for finding the codes in the second table. [Convert the day to a number (Sunday = 1, Monday = 2, etc.). Multiply the number by three.]

Solutions

Date	Code
1	02
2	04
3	06
5	10
12	24
24	48
27	54
30	60

Double the date.

Day	Code
Sunday	03
Monday	06
Tuesday	09
Wednesday	12
Thursday	15
Friday	18
Saturday	21

Convert the day to a number (Sunday = 1, Monday = 2, etc.). Multiply the number by three.

Date	Code
1	104
2	105
3	106
5	107
12	115
24	127
27	130
30	133

Add 103 to the date.

Year: 2005

Date	Code
1	5010
2	5020
3	5030
5	5050
12	5021
24	5042
27	5072
30	5003

Reverse the last two digits of the year (05 becomes 50). Reverse the date (24 becomes 42). Combine the two numbers with the year followed by the date (5042).

* © Copyright 2010. National Governors Association Center for Best Practices and Council of Chief State School Officers. All rights reserved.

Crack the Code

Key Question
What are the missing numbers in the tables of codes?

Learning Goal

Students will:

identify the missing numbers in tables of codes.

Crack the Code

The company *Burgle-tech* has just invented a new "smart alarm." These alarms have codes that change every day of the month (or week) based on a pattern. Codes repeat from month to month (or week to week). You can buy alarms with two-digit codes, three-digit codes, and four-digit codes. Use your problem-solving skills to fill in the missing codes in each of the tables below.

Date	Code
1	02
2	04
3	
5	
12	24
24	
27	
30	60

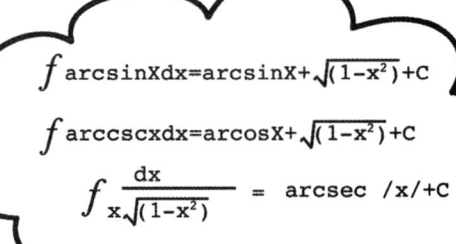

Day	Code
Sunday	03
Monday	
Tuesday	09
Wednesday	
Thursday	
Friday	
Saturday	21

Date	Code
1	104
2	
3	106
5	
12	115
24	127
27	
30	

Year: 2005

Date	Code
1	5010
2	
3	5030
5	
12	5021
24	
27	
30	5003

Use the back of the paper to describe how you figured out the missing codes.

Crack the Code

Connecting Learning

1. What are the missing codes?

2. How were you able to find the missing codes?

3. Which codes were the most difficult to crack? Why do you think this is?

4. What would the code be for the 15th in the first table?

5. What would the code be for today's date in today's year in the last table?

6. Describe the rule for finding the codes in the second table.

Code Consultant

Topic
Problem solving

Key Question
What are the generalizations that make the secret codes work?

Learning Goals
Students will:
- determine the generalizations that lead to a variety of secret codes, and
- create their own codes and generalizations.

Guiding Documents
Project 2061 Benchmark
- Mathematics is the study of many kinds of patterns, including numbers and shapes and operations on them. Sometimes patterns are studied because they help to explain how the world works or how to solve practical problems, sometimes because they are interesting in themselves.

Common Core State Standards for Mathematics*
- Make sense of problems and persevere in solving them. (MP1)
- Model with mathematics. (MP4)
- Look for and make use of structure. (MP7)
- Look for and express regularity in repeated reasoning. (MP8)
- Generate and analyze patterns. (4.OA)

Math
Algebra
    generalizations
Patterns
Problem solving

Integrated Processes
Observing
Comparing and contrasting
Recording
Generalizing

Problem-Solving Strategy
Look for patterns

Materials
Student pages

Background Information
This activity is a continuation of *Crack the Code*, found on the previous pages. In that activity, students are challenged to crack the code for new "smart alarms" invented by the fictitious company *Burgle-Tech*. They have to fill in the missing codes by looking for patterns in the codes they are given. Students will need to have completed *Crack the Code* before they can move on to *Code Consultant*.

While the focus in *Crack the Code* is filling in the missing codes, the focus in this activity is determining the generalizations that lead to the codes and creating new generalizations. The premise for this activity is that a competing company, *Safe-T-Home*, has asked the students to be code consultants. They are to examine several sets of codes used by *Burgle-Tech* and determine the generalizations that make the codes work.

Management
1. Students must have completed *Crack the Code* before doing this activity.
2. The first student page can be done in one math period. The second page should be done the next day or as homework.

Procedure
1. Distribute the first student page and go over the challenge with the class.
2. Allow time for students to work on discovering the generalizations for the codes. If desired, students can work in small groups. Depending on the abilities of your students, these generalizations can be written in words (multiply the date by two), in modified algebraic form (2 x date), or in true algebraic form (2d or 2•d).
3. When students have completed the first page, discuss what they came up with and how they were able to determine the generalizations.
4. Distribute the second student page and allow time for students to develop one or more secret formulas.
5. Have students trade their secret codes with others. Discuss the results.

Connecting Learning
1. What were the secret formulas for the codes? (See *Solutions*.)
2. How did you determine the formulas?
3. Which code was the most difficult to crack? Why?
4. What secret formulas did you come up with?
5. Were your classmates able to discover your formulas? If not, what might you need to change?

Extensions
1. Have students try to come up with some formulas that would make the codes change every month or every year instead of every day. Encourage students to think through what would need to happen in order for the codes to change less frequently.
2. Challenge students to create formulas that would cause the codes to change more than once a day.

Solutions

Table One—Two-digit date code
Double the date and subtract two. (2d – 2)

Table Two—Two-digit day code
Convert days of the week to numbers. Sunday = 1, Monday = 2, etc.
Multiply the number by five and add one. (5n + 1)

Table Three—Three-digit date code
Subtract the date from 500. (500 – d)

Table Four—Four-digit date code
Reverse the digits in the year. Add the date, plus two.

* © Copyright 2010. National Governors Association Center for Best Practices and Council of Chief State School Officers. All rights reserved.

Code Consultant

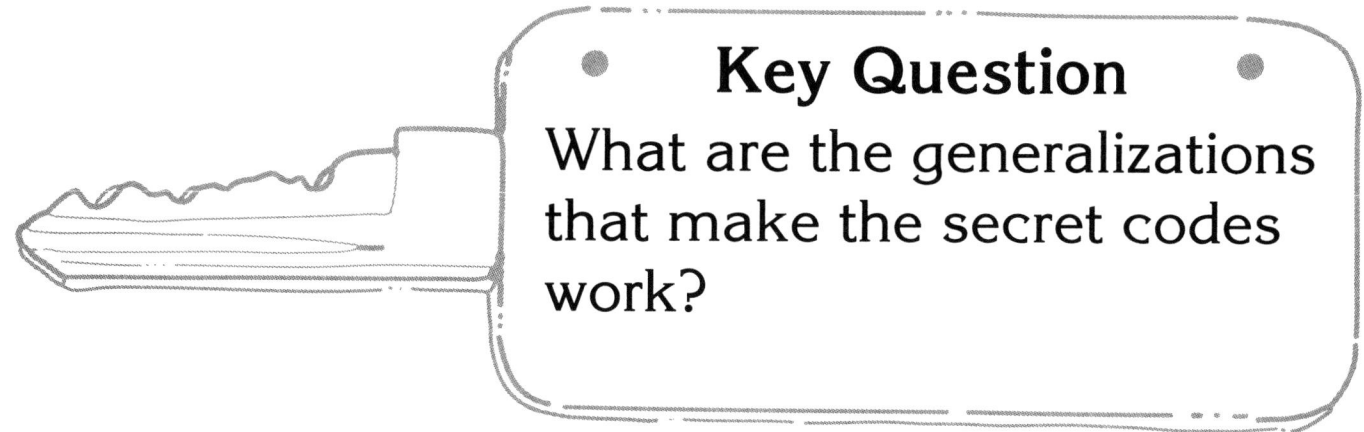

Key Question

What are the generalizations that make the secret codes work?

Learning Goals

Students will:

- determine the generalizations that lead to a variety of secret codes, and

- create their own codes and generalizations.

Code Consultant

Burgle-Tech's "smart alarms" are very popular. They are so popular that another company, Saf-T-Home, has taken the idea. Saf-T-Home is making "intelligent alarms" with codes that change every day based on secret formulas. When you buy one of the alarms, you will get a paper telling you the secret formula so that you can figure out your code each day.

Saf-T-Home wants you to help them discover Burgle-Tech's secret formulas. They also want you to develop some new and better formulas. For each of the code charts below, find the secret formula that Burgle-Tech used to make the code change each day. Write the code in a simple way so that you can plug in any date (or day) and figure out the code.

Date	Code
1	00
2	02
5	08
12	22
24	46
30	58
Secret Formula:	

Day	Code
Sunday	06
Monday	11
Wednesday	21
Friday	31
Saturday	36
Secret Formula:	

Date	Code
1	499
2	498
5	495
12	488
24	476
30	470
Secret Formula:	

Year: 2012

Date	Code
1	2103
2	2104
5	2107
12	2114
24	2126
30	2132
Secret Formula:	

SOLVE IT! 4th

Code Consultant

Now that you are an expert code cracker, it's time to come up with some secret formulas of your own. Fill in the tables below with codes based on your own secret formulas. Make a formula for a two-digit code, a three-digit code, and a four-digit code. When you're finished, trade your paper with a classmate to see if he or she can crack your codes.

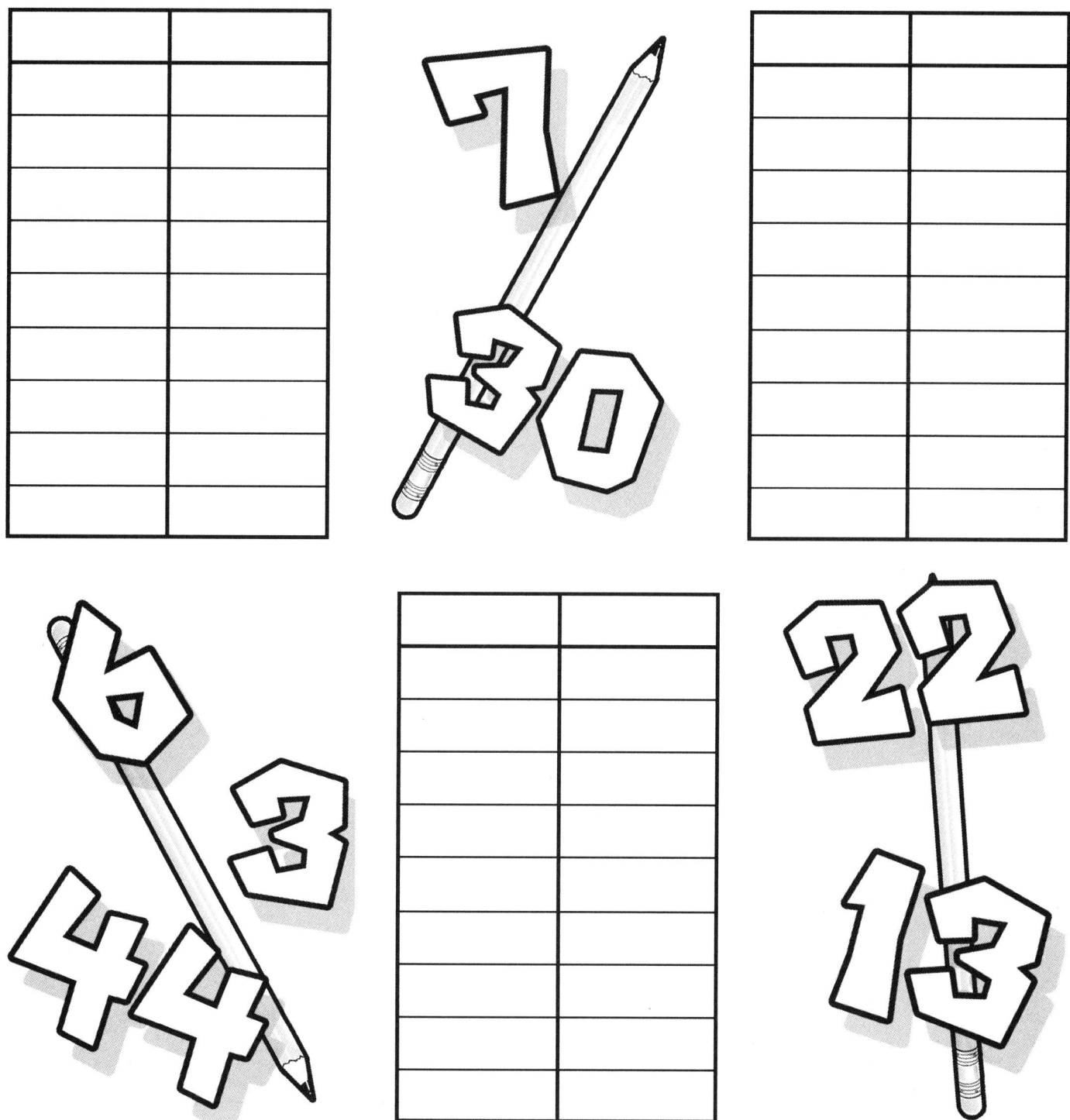

Code Consultant

Connecting Learning

1. What were the secret formulas for the codes?

2. How did you determine the formulas?

3. Which code was the most difficult to crack? Why?

4. What secret formulas did you come up with?

5. Were your classmates able to discover your formulas? If not, what might you need to change?

Tallying the Times Table

Topic
Multiplication

Key Question
How many times does each digit from zero to nine appear in the ones place on a multiplication chart?

Learning Goal
Students will determine how many times each digit from zero to nine appears in the ones place on a multiplication chart.

Guiding Documents
Project 2061 Benchmark
- *Mathematics is the study of many kinds of patterns, including numbers and shapes and operations on them. Sometimes patterns are studied because they help to explain how the world works or how to solve practical problems, sometimes because they are interesting in themselves.*

Common Core State Standards for Mathematics*
- *Make sense of problems and persevere in solving them. (MP1)*
- *Look for and make use of structure. (MP7)*
- *Solve problems involving the four operations, and identify and explain patterns in mathematics. (3.OA)*

Math
Pattern recognition
Number and operations
    multiplication
    odd and even
Problem solving

Integrated Processes
Observing
Comparing and contrasting
Collecting and recording data
Analyzing data

Problem-Solving Strategies
Look for patterns
Organize the information

Materials
Student page

Background Information
If you were to multiply any two whole numbers (0, 1, 2, 3, ...) together at random, what digit is most likely to appear in the ones place in the product? It might seem that all the digits from zero to nine would have an equally likely chance of occurring in the ones place of the product. However, this is not the case. In fact, there is a profound bias toward the even numbers, especially zero. A bit of logical thinking will uncover the reason for this. Whenever two random whole numbers are multiplied together, the first number can be either even or odd and the second number can be even or odd. Thus, there are four possible combinations: an even (number) times an even, an even times an odd, an odd times an even, and an odd times an odd. Whenever one or both of the numbers multiplied is even (which will happen three-fourths of the time according to probability theory), the product is even and therefore the ones digit will be even. Only when both numbers are odd, which only happens one-fourth of the time (according to the laws of probability), will the product end in an odd digit.

This activity, as written, has a narrower scope than the introductory question above. It asks students to consider how many times each digit from zero to nine would appear in the ones place on a multiplication chart. By doing this activity, students will discover the bias towards the even numbers and against the odd numbers.

Management
1. There are two versions of this activity presented—an open-ended one and a more structured one. You will need to decide which one is better suited for your students and distribute the appropriate student page.

Procedure
1. Distribute the appropriate student page and go over the *Key Question.*
2. Allow students time to work on the problem and come up with some extension questions to explore.
3. Have students share their solutions as well as the methods they used to arrive at those solutions.
4. Encourage additional exploration of the extensions that students developed.

Connecting Learning
1. Which number did you guess would be in the ones place most often? Why?
2. Which number is actually in the ones place most often? [zero] Why?

3. What patterns did you notice in your solutions?
4. How can you explain these patterns?
5. What extensions did you think of to explore?
6. What were the results of your explorations?

Extensions
1. Have students determine why zero is the most likely digit to appear in the ones place of a product.
2. Explore what happens when you multiply three digits together instead of two.
3. Look at the probability of having each digit in the ones place of a product.

Solutions

x	0	1	2	3	4	5	6	7	8	9
0	0	0	0	0	0	0	0	0	0	0
1	0	1	2	3	4	5	6	7	8	9
2	0	2	4	6	8	10	12	14	16	18
3	0	3	6	9	12	15	18	21	24	27
4	0	4	8	12	16	20	24	28	32	36
5	0	5	10	15	20	25	30	35	40	45
6	0	6	12	18	24	30	36	42	48	54
7	0	7	14	21	28	35	42	49	56	63
8	0	8	16	24	32	40	48	56	64	72
9	0	9	18	27	36	45	54	63	72	81

Ones Place Digit	0	1	2	3	4	5	6	7	8	9
Frequency	27	4	12	4	12	9	12	4	12	4

- Zero is by far the most likely digit to appear in the ones place of a product: 27 times out of 100.
- The even numbers, not counting zero, each appear 12 times while the odd numbers, with the exception of five, each appear only four times (five appears nine times).
- Since there is a total of 100 products in a zero to nine multiplication table, each of the frequencies produces a percentage (i.e., zero appears in the ones place in 27% of the products, one appears in 4%, two appears in 12%, etc.).
- When looking at totals for the odd and even numbers, there are 75 even digits in the ones place and 25 odd.

* © Copyright 2010. National Governors Association Center for Best Practices and Council of Chief State School Officers. All rights reserved.

Tallying the Times Table

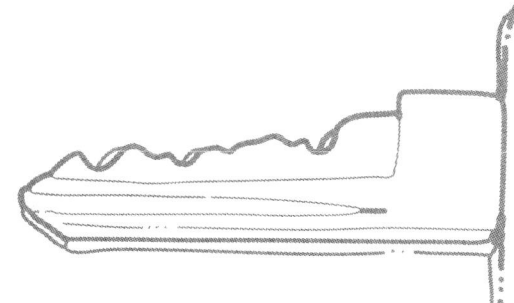

Key Question

How many times does each digit from zero to nine appear in the ones place on a multiplication chart?

Learning Goal

Students will:

determine how many times each digit from zero to nine appears in the ones place on a multiplication chart.

Tallying the Times Table
Structured Version

What digit appears most frequently in the ones place in a zero to nine multiplication table? Make a guess—then fill in the charts to help you find the answer. It will help to underline the ones digit in each product in the multiplication table as shown in the five times row.

x	0	1	2	3	4	5	6	7	8	9
0										
1										
2										
3										
4										
5	0	5	10	15	20	25	30	35	40	45
6										
7										
8										
9										

Ones Place Digit	0	1	2	3	4	5	6	7	8	9
Frequency										

Think of some questions that are related to this problem. Pick one for further exploration and report your findings on the back of this paper.

Tallying the Times Table
Open-Ended Version

What digit appears most frequently in the ones place in a zero to nine multiplication table? Make a guess and then find out if you are correct. Show the process you used to answer this question and any discoveries you make in the space below.

My guess:

Think of some other questions related to this problem. Pick one to explore and report your findings on the back of this paper.

Tallying the Times Table

Connecting Learning

1. Which number did you guess would be in the ones place most often? Why?

2. Which number is actually in the ones place most often? Why?

3. What patterns did you notice in your solutions?

4. How can you explain these patterns?

5. What extensions did you think of to explore?

6. What were the results of your explorations?

Problem-Solving Strategies
Use Logical Thinking

Sometimes a problem requires logical thinking. You must find an answer when you are missing some of the information. You must make inferences based on what you know. Often logic problems have many clues. Sometimes grids are used to organize the clues. Other times manipulatives are used.

MADE-TO-ORDER RECTANGLES

Topic
Logical thinking

Key Question
How can you fill the square grids with rectangles according to certain rules?

Learning Goal
Students will use logical thinking to solve number grid problems.

Guiding Document
Common Core State Standards for Mathematics*
- *Make sense of problems and persevere in solving them. (MP1)*

Math
Number sense
One-to-one correspondence
Geometry
    rectangular shapes

Problem-Solving Strategies
Use logical thinking
Use manipulatives
Guess and check

Integrated Processes
Observing
Comparing and contrasting
Collecting and recording data
Analyzing
Applying

Materials
Area Tiles, 10 per student
Student pages

Background Information
    This activity is a simplified and extended version of a puzzle that appeared in the September 2004 *Games* magazine. The original consisted of large square grids (10 x 10 to 10 x 18) that were partially filled with numbers ranging from 2 to 17. The challenge was to enclose the numbers with rectangles that were made up of the number of small squares equivalent to the value of the number. For example, the number 5 would be in a five-square rectangle, the number 6 would be in a six-square rectangle, and so on. Every square on the grid had to be part of a rectangle, and no rectangles could overlap. The numbers were placed in the grids in such a way that there were usually multiple possible rectangles that could be created for each number. In order to solve the puzzles, you had to find the one number that could only be contained by one possible rectangle and work your way out from there.

    Our version presents much smaller grids (the largest is 7 x 7), and begins by giving students experience with combining different numbers of square tiles to form all the possible rectangular shapes. This introductory experience will make them familiar with all of the shapes they will be using to solve the problems, allowing them to focus on the logic and critical thinking rather than the mechanics.

Management
1. Each student will need 10 Area Tiles to manipulate during the first part of this activity. Area Tiles are available from AIMS (item number 4810). If Area Tiles are not available, other uniform square manipulatives can be used instead.
2. This activity is divided into two parts. The first part can be done either individually or in collaborative groups. Students should do the second part individually.
3. There are three student pages for *Part Two*. The problems increase in difficulty as the grids get larger. Give students the page(s) appropriate for their ages and abilities.

Procedure
Part One
1. Distribute the first student page and 10 Area Tiles to each student.
2. Go over the instructions as a class and be sure that everyone understands what they are supposed to do.
3. Provide sufficient time for students to work with the Area Tiles and discover all of the possible rectangular arrangements for two through 10 tiles.
4. Encourage students to record both horizontal and vertical arrangements.

Part Two
1. Distribute the student page(s) for *Part Two* and explain the challenge. If necessary, complete one problem on the overhead so that students can get an idea of how to go about solving them.

SOLVE IT! 4th   © 2012 AIMS Education Foundation

2. Give students time to work on the problems and come up with solutions. If they need manipulatives to help them with the solution process, provide different colors of Area Tiles with which to make the rectangles.
3. Close with a time of class discussion where students share their strategies and techniques for solving the problems.

Connecting Learning

Part One
1. How many ways can you put two Area Tiles together to make a rectangle? [one; two if you count each orientation as a different rectangle] …three Area Tiles? [one/two] …four Area Tiles, etc.?
2. Which number(s) of Area Tiles had the most rectangular arrangements? [six and eight] Why do you think this is?
3. Which number(s) of Area Tiles had the fewest rectangular arrangements? [three, five, and seven] Why do you think this is? [They are odd numbers and not squares, so they can only form one kind of rectangle.]
4. Which numbers of tiles could form squares? [four and nine] Why? [They are square numbers.]

Part Two
1. How did you first approach the problems?
2. What strategies did you develop that helped you solve each one?
3. Did your approach change after you had solved a couple? Why or why not?
4. Did you discover any "unwritten rules" as you solved the problems?
5. If you could give one helpful hint to someone who has never seen these problems before, what would it be?

Extensions
1. Make up larger, more complicated grids for students to solve.
2. Have students create their own grids and trade them with classmates to solve.

Solutions

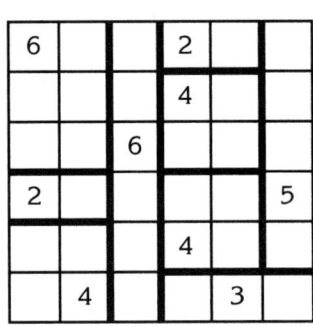

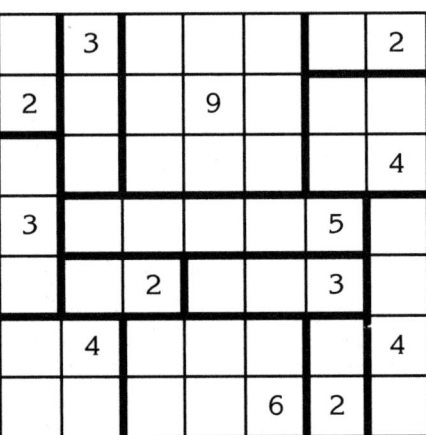

* © Copyright 2010. National Governors Association Center for Best Practices and Council of Chief State School Officers. All rights reserved.

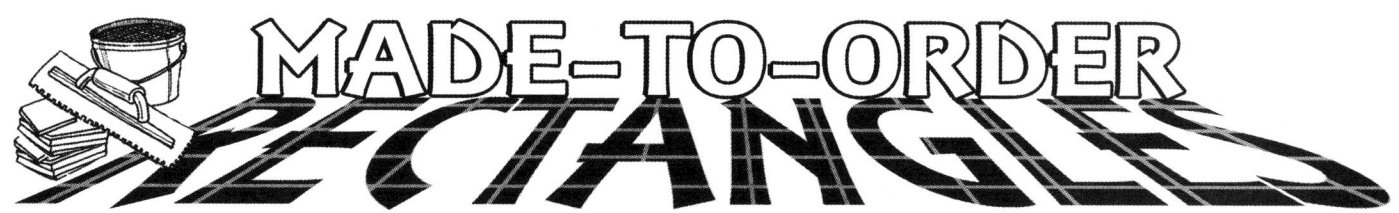

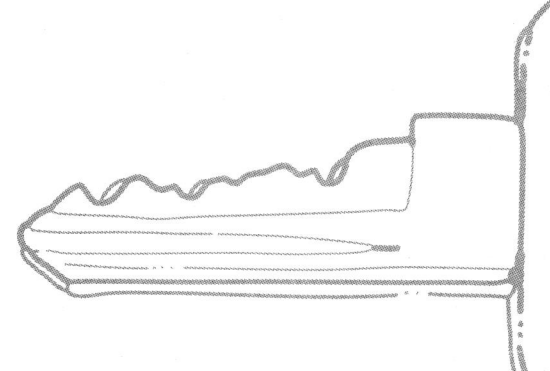

Key Question

How can you fill the square grids with rectangles according to certain rules?

Learning Goal

use logical thinking to solve number grid problems.

Use different numbers of Area Tiles to make rectangles. Record every different rectangle you can make using two through 10 Area Tiles. (Don't forget that squares are also rectangles.)

MADE-TO-ORDER RECTANGLES

Sample Problem

Fill each large square with rectangles so that the following statements are true:
- Each rectangle contains one number.
- The number in each rectangle is equal to the number of small squares that make up that rectangle
- Every small square is part of a rectangle.
- No rectangles overlap.

Don't forget that squares are also rectangles!

SOLVE IT! 4th   201   © 2012 AIMS Education Foundation

MADE-TO-ORDER RECTANGLES

Fill each large square with rectangles so that the following statements are true:
- Each rectangle contains one number.
- The number in each rectangle is equal to the number of small squares that make up that rectangle.
- Every small square is part of a rectangle.
- No rectangles overlap.

Don't forget that squares are also rectangles!

Sample Problem

SOLVE IT! 4th   202   © 2012 AIMS Education Foundation

MADE-TO-ORDER RECTANGLES

Fill each large square with rectangles so that the following statements are true:

- Each rectangle contains one number.
- The number in each rectangle is equal to the number of small squares that make up that rectangle.
- Every small square is part of a rectangle.
- No rectangles overlap. Don't forget that squares are also rectangles!

Sample Problem

			3
2		4	
2			
		3	2

			3
2		4	
2			
		3	2

	3					2
2			9			
						4
3					5	
		2			3	
	4					4
					6	2

SOLVE IT! 4th   203   © 2012 AIMS Education Foundation

MADE-TO-ORDER RECTANGLES

Connecting Learning

Part One

1. How many ways can you put two Area Tiles together to make a rectangle? …three Area Tiles? …four Area Tiles, etc.?

2. Which number(s) of Area Tiles had the most rectangular arrangements? Why do you think this is?

3. Which number(s) of Area Tiles had the fewest rectangular arrangements? Why do you think this is?

4. Which numbers of tiles could form squares? Why?

Connecting Learning

Part Two

1. How did you first approach the problems?

2. What strategies did you develop that helped you solve each one?

3. Did your approach change after you had solved a couple? Why or why not?

4. Did you discover any "unwritten rules" as you solved the problems?

5. If you could give one helpful hint to someone who has never seen these problems before, what would it be?

Shape Logic

Topic
Logical thinking

Key Question
How can you place numbers into the given shapes so that the statements below the shapes are true?

Learning Goal
Students will use logical thinking to place numbers in a variety of sub-divided shapes so that the sets of clues describing the shapes are true.

Guiding Documents
Project 2061 Benchmarks
- *Mathematical statements using symbols may be true only when the symbols are replaced by certain numbers.*
- *Add, subtract, multiply, and divide whole numbers mentally, on paper, and with a calculator.*

Common Core State Standards for Mathematics*
- *Make sense of problems and persevere in solving them. (MP1)*

Math
Whole number operations
Number sense
    even/odd
    greater than/less than
Logic
Problem solving

Integrated Processes
Observing
Comparing and contrasting
Recording

Problem-Solving Strategies
Use logical thinking
Use manipulatives
Guess and check

Materials
Student pages
Number cards (see *Management 2*)

Background Information
This activity gives students the opportunity to develop their logical thinking skills by filling squares, hexagons, and octagons with numbers based on clues. The clues use the language of mathematics, giving students the opportunity to practice their mathematical vocabulary as well as some basic computation.

Management
1. Depending on the abilities of your students, you may wish to use only one or two of the student pages provided. The first student page has square grids in which students are challenged to place either the numbers one through four or five through eight. The second student page has hexagons and an octagon, in which students are challenged to place the numbers one through six and one through eight, respectively. The final student page provides a blank square, hexagon, and octagon in which students are challenged to create their own logic problems to share with their classmates.
2. Have students cut out small number cards that will fit into the spaces on the student sheets so they have something to manipulate as they try to determine the solution for each problem.

Procedure
1. Distribute the first student page and go over the challenge with the class.
2. Have students create number cards to use while doing the problems.
3. Allow time for students to complete the problems on the first page before giving them the second page.
4. The final student page can serve as an assessment, if desired. Have students record the solution to each puzzle they create on a separate sheet of paper, and write the clues to determine each solution on the student page. (The number of clues may be greater or less than the number of lines provided.) A puzzle can be judged successful if it can be solved using only the clues given. Encourage students to think of clues that use mathematical concepts and principles.
5. Conclude the activity with a time of discussion in which students share their strategies for determining solutions as well as creating their own puzzles.

SOLVE IT! 4th   © 2012 AIMS Education Foundation

Connecting Learning

1. How were you able to solve the problems?
2. Were some more difficult than others? Which ones? Why?
3. Did having the number cards help you solve the problems? How?
4. What process did you go through to create your own problems?
5. Were your problems successful? How do you know?

Solutions

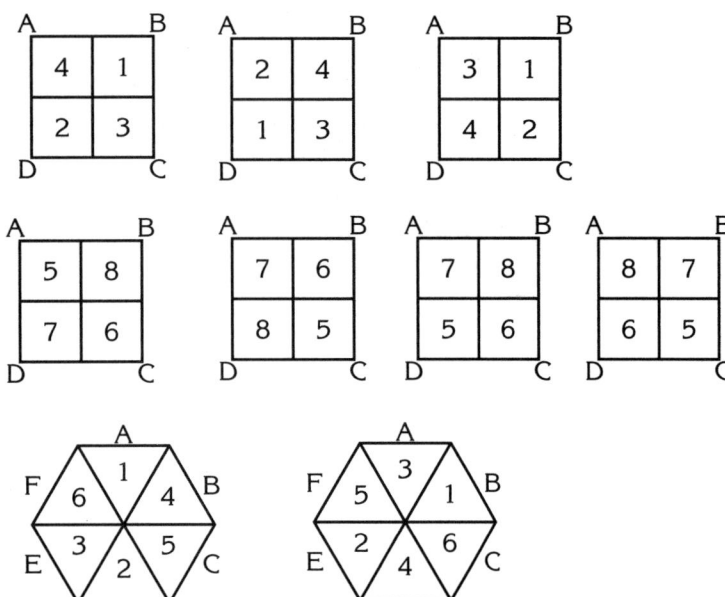

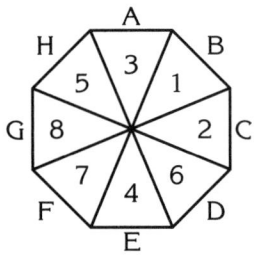

* © Copyright 2010. National Governors Association Center for Best Practices and Council of Chief State School Officers. All rights reserved.

Key Question

How can you place numbers into the given shapes so that the statements below the shapes are true?

Learning Goal

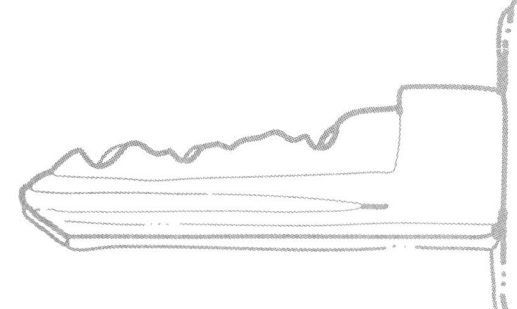

use logical thinking to place numbers in a variety of sub-divided shapes so that the sets of clues describing the shapes are true.

Shape Logic — Square Logic

Use the clues provided to fill in each square with the numbers one through four.

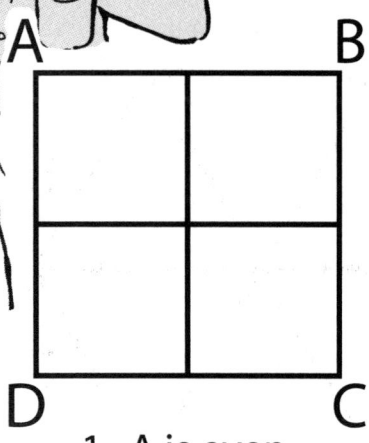

1. A is even
2. A − 3 = B
3. C > D

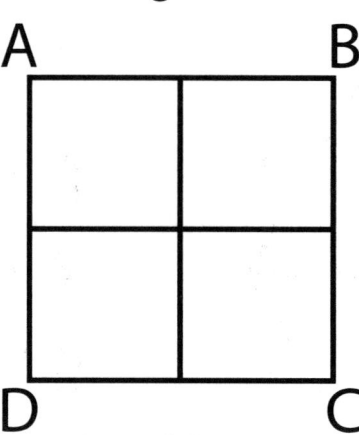

1. A and B are even
2. D < C
3. B > C

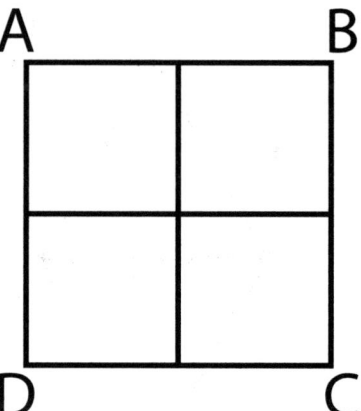

1. B + D = A + C
2. A and B are odd
3. A > C

Use the clues provided to fill in each square with the numbers five through eight. The last problem has two solutions. See if you can find them both.

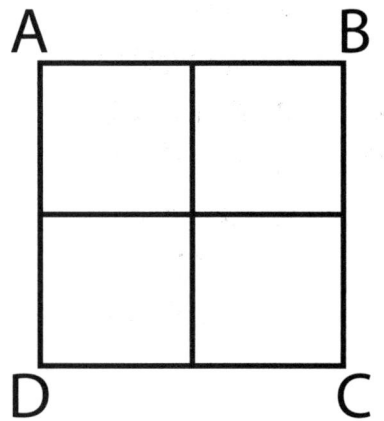

1. A + C = 11
2. B + C = 14

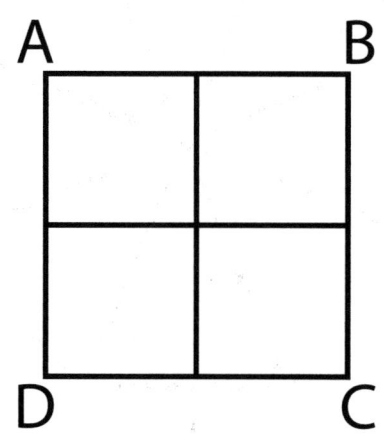

1. B + D = A x 2
2. B + C = 11

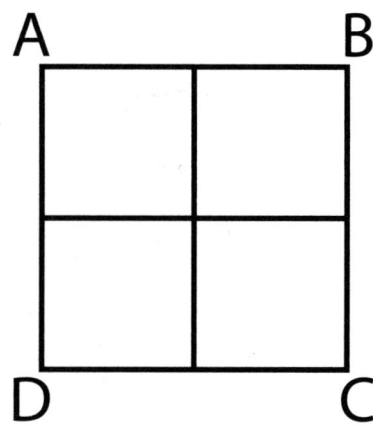

1. B + D = A + C
2. A + B = 15

SOLVE IT! 4th © 2012 AIMS Education Foundation

Shape Logic — Hexagon Octagon Logic

Use the clues provided to fill in each hexagon with the numbers one through six and the octagon with the numbers one through eight.

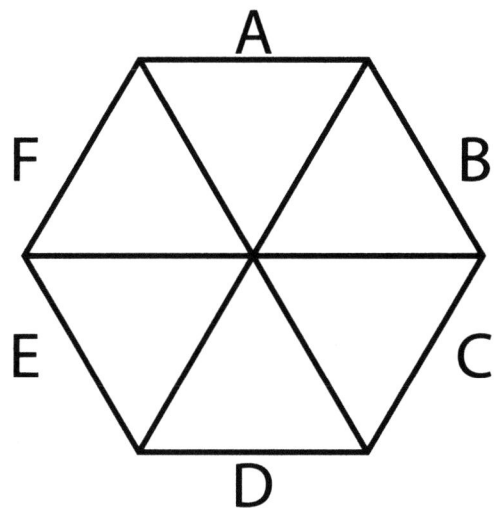

1. A + 5 = F
2. F ÷ 2 = E
3. B + C = F + E
4. C > B

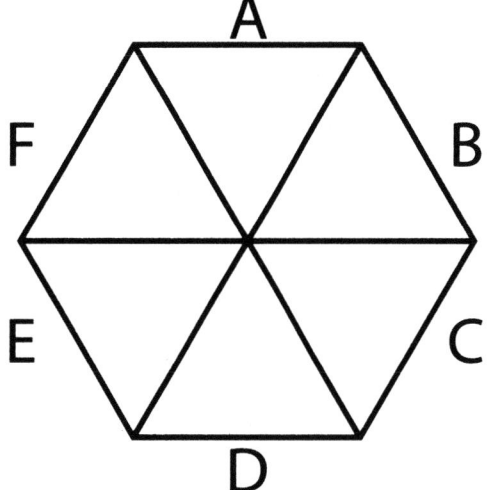

1. F x B = 5
2. A + B = 4
3. 2 x A = C
4. E x 2 = D

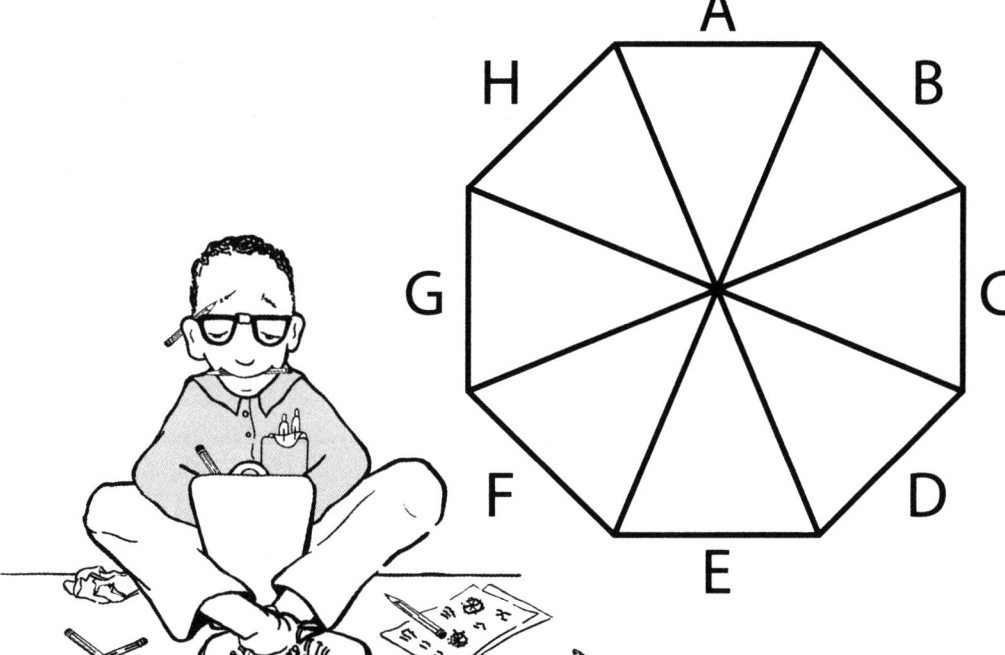

1. B + C = A
2. B = 1
3. C is even
4. A x 2 = D
5. E = C x C
6. G – B = F
7. C x E = G
8. E + B = H

SOLVE IT! 4th © 2012 AIMS Education Foundation

Use the empty spaces to create your own logic problems and then trade them with a classmate. Remember, the goal is not to stump your classmates, but to create puzzles that they can solve.

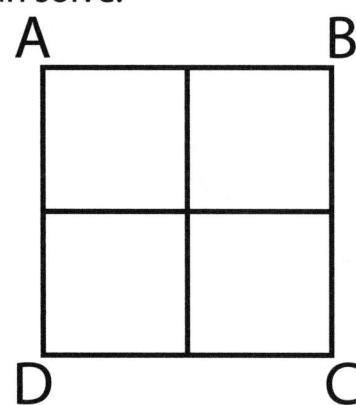

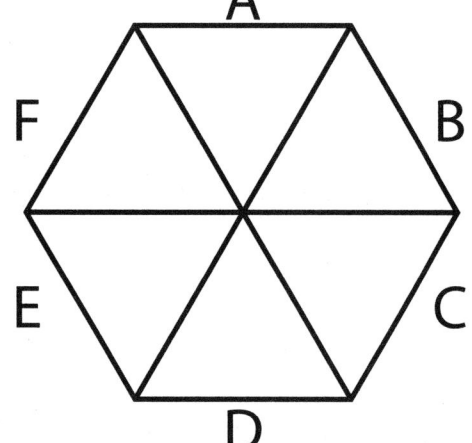

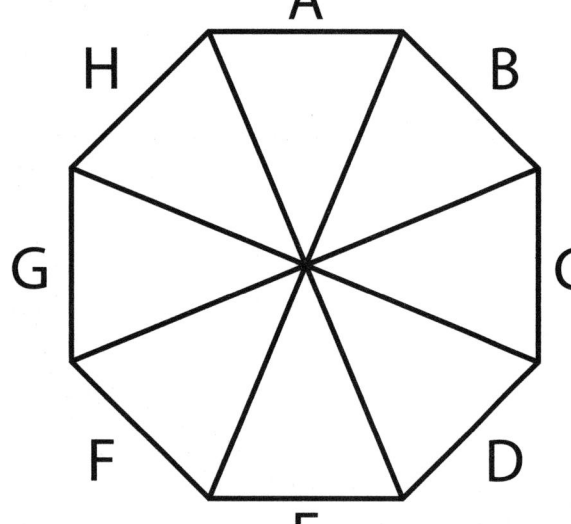

SOLVE IT! 4th   212   © 2012 AIMS Education Foundation

Shape Logic

Connecting Learning

1. How were you able to solve the problems?

2. Were some more difficult than others? Which ones? Why?

3. Did having the number cards help you solve the problems? How?

4. What process did you go through to create your own problems?

5. Were your problems successful? How do you know?

Nabbing Numbers

Topic
Logical thinking

Key Question
What strategies can you develop to end the game with the most points?

Learning Goal
Students will play a game that requires them to use logical thinking and develop strategies in order to end with the most points.

Guiding Document
Common Core State Standards for Mathematics*
- Make sense of problems and persevere in solving them. (MP1)

Math
Number and operations
    addition
    integers
Logical thinking
Problem solving

Integrated Processes
Observing
Recording

Problem-Solving Strategies
Use logical thinking
Use manipulatives

Materials
For each pair of students:
    game board
    number cards, one set
    scissors

For each student:
    student page

Background Information
This activity is an original game that helps students develop their critical thinking skills and gives them practice in basic computation. In this game, the pieces move like the knight in a game of chess. Starting on a square of one color, an L-shaped pattern is followed to end on a square of a different color. The following diagram shows all of the possible squares on which the black star could land.

The object of the game is to collect number cards worth the most points. Every time a player lands on a space containing a number card, that card is captured, and it is removed from the board. However, points are only received for number cards that are the player's color. In other words, the gray player only gets points for gray cards, and the white player only gets points for white cards.

The game ends when all of the cards of one color have been captured, regardless of the number of other cards remaining on the board. It is likely that players will end up with different numbers of cards, because it is not always possible to capture a number card every turn.

Management
1. This game is designed to be played in pairs. Each pair of students will need one game board and one set of the appropriate number cards. (Two sets of number cards have been put on one page to save paper.) Both pages should be copied onto cardstock for durability and for ease of handling.
2. To set up the game, the number cards must be placed in the squares on the grid so that the color of the number card corresponds to the color of the square it is in. The cards should be mixed up and placed on the board randomly, not in numerical order, although cards of the same value in different colors may be next to each other. The squares labeled *white start* and *gray start* are for the starred pieces that serve as the playing chips. The white player always begins, so you may want to have students trade colors for each round they play.

SOLVE IT! 4ᵗʰ   215   © 2012 AIMS Education Foundation

Procedure

1. Have students get into pairs and distribute the student page, a game board, and one set of number cards. Instruct students to cut apart the number cards and go over how to set up the board (see *Management 2*).
2. Read through the rules as a class and make sure that everyone understands how to play and the way the pieces move. Have players pick their colors.
3. Allow time for students to play a round. Once the round ends, have students total their points for that round to determine the winner. Have students keep track of their scores on a sheet of scratch paper for later discussion.
4. Have students mix up the number cards and place them on the grid again, in a different order. Instruct them to switch colors before playing another round.
5. Play as many rounds as time allows before having students answer the questions on the student page and holding a closing discussion. Challenge students to articulate their strategy(ies) and determine if one strategy seemed to dominate.

Connecting Learning

1. What strategies did you develop as you played this game?
2. Did you find yourself planning ahead and trying to position yourself for moves one or two turns away?
3. Did you focus on going after your own points, or did you try to keep the other player from capturing points?
4. Which strategy(ies) seemed to be the most effective?
5. Did one player out of each pair seem to win more than the other?
6. What factors could account for this?

Extension

For students who are ready to deal with integers, a twist can be added after several rounds of the original game have been played. Instead of getting no points for cards of the other player's color, negative points are received. That is, the white player gets positive points for each white card, but negative points for each gray card. This significantly changes the strategy necessary, and adds more of a computational challenge as well.

* © Copyright 2010. National Governors Association Center for Best Practices and Council of Chief State School Officers. All rights reserved.

SOLVE IT! 4th 216 © 2012 AIMS Education Foundation

Nabbing Numbers

Key Question

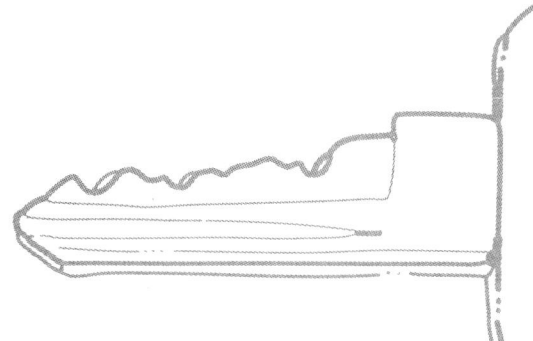

What strategies can you develop to end the game with the most points?

Learning Goal

Students will:

play a game that requires them to use logical thinking and develop strategies in order to end with the most points.

 **Number Cards**

16	16	14	14	13	13	12	12
11	11	10	7	9	9	8	8
7	7	6	6	5	5	★	★
							3

Cut out these number cards and put them on your grid in random order.

 ✂ ━━━━━━━━━━━━━━━━━━━━━━━━━━━━━━━━━━━━━━━▶

Cut out these number cards and put them on your grid in random order.

16	16	14	14	13	13	12	12
11	11	10	7	9	9	8	8
7	7	6	6	5	5	★	★
							3

SOLVE IT! 4ᵗʰ 218 © 2012 AIMS Education Foundation

  **Game Board**

				white start
				gray start

Nabbing Numbers

Setup
Once you have cut out your number cards, lay them face up on the game board in random order. White cards go on white spaces, and gray cards go on gray spaces. The two starred pieces are the playing chips and go on the spots marked **white start** and **gray start.** The white player begins first. Playing chips move like the knight in a game of chess—they follow an L-shaped pattern to end on a square of the opposite color. The grid below shows all of the possible moves the black star could make.

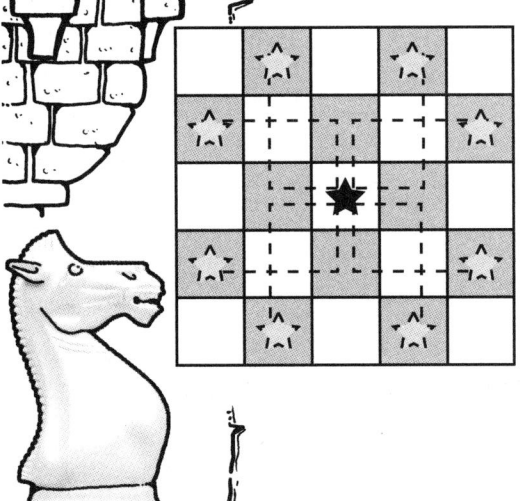

Rules
The object of the game is to get the most points. Each time you land on a space with a number card, you capture that card. However, you only receive points for cards that are the same color as your game piece. In other words, the white player only gets points for white cards, and the gray player only gets points for gray cards. The game ends when all of the cards of one color have been removed from the board, no matter how many cards of the other color are left. Total your points to determine the winner, then shuffle the number cards, lay them out, and play again. Keep track of your scores for each round on a piece of scratch paper.

Answer these questions after you have played several rounds with your partner. Be prepared to share your answers with the class.

1. Describe the strategy you used the first time you played the game.

2. Did this strategy change as you played more rounds? Why or why not?

3. Based on how often you won, was your strategy effective?

Connecting Learning

1. What strategies did you develop as you played this game?

2. Did you find yourself planning ahead and trying to position yourself for moves one or two turns away?

3. Did you focus on going after your own points, or did you try to keep the other player from capturing points?

4. Which strategy(ies) seemed to be the most effective?

5. Did one player out of each pair seem to win more than the other?

6. What factors could account for this?

Problem-Solving Strategies
Wish for an Easier Problem

Sometimes a problem has lots of data or big numbers. It can seem too hard to do. This is when you can "wish for an easier problem." You can use smaller numbers instead of the big ones. You can think about how to solve the problem instead of the numbers and data. This will help you see how to solve a simpler version of a harder problem.

Our Library in the Limelight

Topic
Problem solving

Key Question
How many words are in the library book you are currently reading?

Learning Goals
Students will:
- make a guess about the number of words in the book they are currently reading,
- devise a method to come up with a more accurate estimate without having to count each word in the book,
- brainstorm additional questions related to the library, and
- pick one of these questions to explore in depth.

Guiding Document
Common Core State Standards for Mathematics*
- *Make sense of problems and persevere in solving them. (MP1)*

Math
Estimation
Problem solving

Integrated Processes
Observing
Predicting
Collecting and recording data
Organizing data
Generalizing
Applying

Problem-Solving Strategies
Wish for an easier problem
Organize the information

Materials
Library access
Student page

Background Information
National Library Week is celebrated annually during the third week in April. *Our Library in the Limelight* is designed to be used in conjunction with this important event.

This activity has much potential *if* students can come up with good questions to research. Thus, the brainstorming process is critical. During this time, your facilitation (to help students come up with good questions) can make or break the activity. After students have generated many questions, they need to be encouraged to pick one they can explore in depth.

Management
1. This activity has two parts. The first part is done in class, while the second part requires time spent in the library. The second part can take place in groups or individually, but should result in a report that is presented to the whole class at a later time. This information can also be presented in the form of a poster that could be put up in the classroom or library during National Library Week.
2. Doing this activity will require the assistance of your librarian. Make sure you coordinate this ahead of time and arrange for your students to meet with the librarian and/or spend time in the library to do their research.

Procedure
1. Ask the *Key Question*. Have students share their guesses.
2. Distribute the student page and have students record their guesses.
3. Allow them to work in small groups or individually to brainstorm ways to make a more accurate estimation.
4. Have them compare their original guesses to the estimates they got when they used a more organized approach.
5. After sharing responses, have students and/or groups brainstorm a variety of questions that are related to their library and then pick one question to explore in depth. Some possible questions follow:
   - How many books are in the library?
   - Which class checks out the most books?
   - How many books are in circulation each week?
   - What is the ratio of books in circulation to books on the shelves?
   - Of which type of book does the library have more—fiction or non-fiction?
   - Which of these is more popular?
   - What is the most used section in the library?
   - How long does the average library book last?
   - What is the oldest book in the library? ...the newest?
   - How many library books are read by each grade level each week? ...each year?

- On what day of the week are the most books checked out? ...checked in?
- What is the average cost for a new library book?
- What is the library budget for new books?
- How many books can be bought each year?

6. Provide time over a period of days or weeks for students to research their questions. Determine the format that you would like their final reports to take, and have a day where these reports are shared.

Connecting Learning
1. What method did you develop to determine the number of words in your library book?
2. When you applied this method, how did your new estimate compare to your original guess?
3. Which of the methods used do you think gives the best estimate? Why?
4. What other questions did you think of to explore?
5. What will you need to do to find an answer to this question?
6. How will what you learned in the first part of the activity help you?

* Reprinted with permission from *Principles and Standards for School Mathematics*, 2000 by the National Council of Teachers of Mathematics. All rights reserved.

Our Library in the Limelight

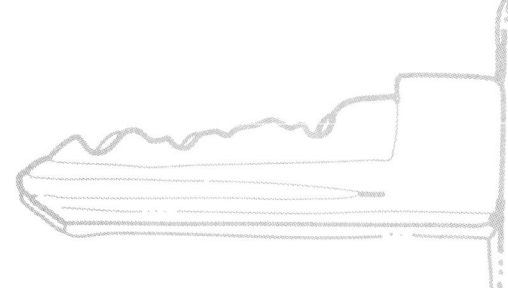

Key Question

How many words are in the library book you are currently reading?

Learning Goals

Students will:

- make a guess about the number of words in the book they are currently reading,

- devise a method to come up with a more accurate estimate without having to count each word in the book,

- brainstorm additional questions related to the library, and

- pick one of these questions to explore in depth.

Our Library in the Limelight

How many words are in your library book? Make a guess.

How could you come up with a more accurate estimate without counting each word in the book? Describe the process you would use and then apply it. Compare your estimate with your original guess.

Brainstorm some questions about your library and write them below.

Pick one or more of these questions to explore in depth. Devise a plan for finding the answer to your question and then make a report of your findings. Be prepared to share your report with the rest of the class.

Our Library in the Limelight

Connecting Learning

1. What method did you develop to determine the number of words in your library book?

2. When you applied this method, how did your new estimate compare to your original guess?

3. Which of the methods used do you think gives the best estimate? Why?

4. What other questions did you think of to explore?

5. What will you need to do to find an answer to this question?

6. How will what you learned in the first part of the activity help you?

Calculating a Classroom Crunch

Topic
Problem solving

Key Question
If all the furniture were removed, how many students would fit in your classroom, with each student standing on the floor?

Learning Goals
Students will:
- brainstorm ways to answer the *Key Question*,
- select one of the methods and describe it in detail, and
- put this plan into action to determine an estimate of the number of students who could fit into the classroom.

Guiding Document
Common Core State Standards for Mathematics*
- *Make sense of problems and persevere in solving them. (MP1)*
- *Use the four operations with whole numbers to solve problems. (4.OA)*

Math
Estimation
Problem solving

Integrated Processes
Observing
Comparing and contrasting
Collecting and recording data
Organizing data
Generalizing
Applying

Problem-Solving Strategies
Wish for an easier problem
Organize the information

Materials
Student pages
Requested student materials (see *Management 1*)

Background Information
This activity is an open-ended exercise that focuses on the process of problem solving rather than on getting the correct answer—which in this particular problem is unknown. The question posed is a simple one: If all the furniture were removed, how many students would fit in your classroom, with each student standing on the floor?

This exercise has three distinct parts, each of which is important. The first part is a brainstorming session that challenges each group to come up with as many different ways to answer the question as they can. The second part of this exercise has groups putting their plans into action. The third part of the exercise is a whole-class discussion. During this time students should share their methods and results. The discussion should make it obvious that some results are more accurate than others and that some methods are easier than others. By comparing the results, the class should be able to reach a fairly accurate estimate of the number of students that would fit in the room.

Management
1. If some groups develop plans that require the use of meter sticks or measuring tapes, these items should be made available.
2. Groups of four to six students work well.

Procedure
1. Ask the *Key Question* and have students make some guesses.
2. Divide students into groups and distribute the first student page.
3. Provide time for groups to brainstorm and record some methods for answering the question. Encourage them to pick the method they would like to use and describe it in detail, knowing that they will be putting this plan into action.
4. Remind groups that they will not be able to actually remove all the furniture and cram students into the classroom since this would be against the fire code, as well as being dangerous and impractical. Groups that come up with a workable plan before others have finished their plans should be encouraged to come up with an alternate plan.
5. Once all groups have developed a plan, provide a time for them to begin putting their plans into action. As groups experiment with their various techniques, those with marginally effective plans will discover that there may be better ways to find the answer.
6. Distribute the second student page and have groups record their results.
7. When all groups have had time to develop an estimate, have students share their methods and results. Give students the final student page and have them respond to the questions.

SOLVE IT! 4th © 2012 AIMS Education Foundation

8. Determine as a class a fairly accurate estimate of the number of students that would fit in the room.

Connecting Learning
1. What method did your group come up with for finding the number of people who could fit in your classroom?
2. Was your plan easy to carry out? Why or why not?
3. How does your method compare to the methods used by other groups?
4. Which method do you think is the most accurate? Why do you think this is?
5. Which method is the easiest to put into action? Why?
6. Which method do you like best? Why?
7. Based on all of the results, what is a good estimate of the number of students that would fit in our classroom?

Extensions
1. How would the answer change if students had to space themselves so they were at least an arm's length from every other student in the room?
2. How many students would fit in the classroom if they were lying on the floor instead of standing?
3. How many students would fit if people could sit on the shoulders of other students?
4. How many students would fit if you used first graders? …high schoolers?

Solutions
This problem can be solved in a number of different ways. The following suggestions offer a few of those ways, but your students may come up with others.

Corner Cram
Have the people in the group cram into a corner of the room and measure the floor space they occupy. Find the total area of the floor and divide it by the area taken up by the group. This result is multiplied by the number of people in the group to come up with an answer.

Tile Count
In a room with floor tiles, determine how many students can stand on every two, three, or four tiles. Count the number of tiles in the room, divide by two (or three or four) and multiply by the number of students that can stand on that number of tiles. A similar method counts the number of students who can stand beneath one ceiling tile and multiplies this number by the total number of tiles.

Leap Frog
Line the group up shoulder to shoulder and then "leap frog" across the width of the room by having the person at the end of the line move to the beginning until the group has traveled across the room. Repeat with students standing front to back and moving across the length of the room. Multiply the two numbers together to determine the total number of students who could fit in the room.

* © Copyright 2010. National Governors Association Center for Best Practices and Council of Chief State School Officers. All rights reserved.

Calculating a Classroom Crunch

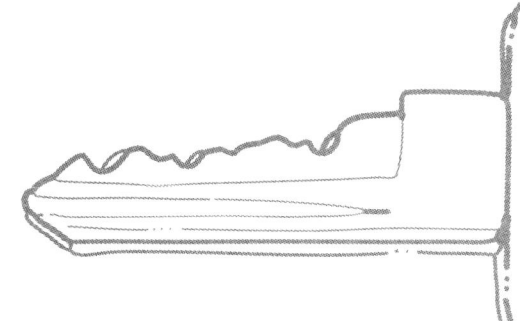

Key Question

If all the furniture were removed, how many students would fit in your classroom, with each student standing on the floor?

Learning Goals

Students will:

- brainstorm ways to answer the *Key Question*,

- select one of the methods and describe it in detail, and

- put this plan into action to determine an estimate of the number of students who could fit into the classroom.

Calculating a Classroom Crunch

If all the furniture were removed, how many students would fit in your classroom, with each student standing on the floor?

Part One
With your group, brainstorm several ways to find the answer to this question and list them below.

Pick one of these methods that does not require removing all the furniture and cramming students into the classroom. (This is a violation of the fire code.) You will use your chosen method to come up with an answer to the question. Describe your plan in detail.

Calculating a Classroom Crunch

If all the furniture were removed, how many students would fit in your classroom, with each student standing on the floor?

Part Two
Put your plan from *Part One* into action. Describe your results and the process(es) you used to come up with these results below.

Calculating a Classroom Crunch

If all the furniture were removed, how many students would fit in your classroom, with each student standing on the floor?

Part Three

As a class, share the results achieved and processes used by the various groups, then answer the following questions.

1. Which method do you think is the most accurate? Why?

2. Which method is the easiest to put into action? Why?

3. Which method do you like best? Why?

4. What extensions can you think of for this problem?

Calculating a Classroom Crunch

Connecting Learning

1. What method did your group come up with for finding the number of people who could fit in your classroom?

2. Was your plan easy to carry out? Why or why not?

3. How does your method compare to the methods used by other groups?

4. Which method do you think is the most accurate? Why?

5. Which method is the easiest to put into action? Why?

6. Which method do you like best? Why?

7. Based on all of the results, what is a good estimate of the number of students that would fit in our classroom?

A Close Call

Topic
Estimation

Key Question
How close can you estimate the number of objects in the bag?

Learning Goals
Students will:
- estimate the number of objects in a container, and
- devise a strategy for calculating a close approximation to the exact number without counting each one.

Guiding Document
Common Core State Standards for Mathematics*
- *Make sense of problems and persevere in solving them. (MP1)*
- *Reason abstractly and quantitatively. (MP2)*
- *Construct viable arguments and critique the reasoning of others. (MP3)*
- *Model with mathematics. (MP4)*
- *Use appropriate tools strategically. (MP5)*
- *Use the four operations with whole numbers to solve problems. (4.OA)*

Math
Estimation
Data analysis
  averages
Problem solving

Integrated Processes
Observing
Comparing and contrasting
Recording data
Generalizing

Problem-Solving Strategies
Wish for an easier problem
Work backwards
Organize the information

Materials
For each group:
  1 zipper-type plastic bag, pint size
  objects (see *Management 1*)
  balance
  metric masses
  small measuring cup

Background Information
This activity is designed to strengthen the student's ability to communicate effectively in words and symbols what has been experienced in the real world in a problem-solving situation. The "doing" process is relatively easy, but matching the proper words to the process is difficult and the connecting the action to the algorithm is most difficult but very powerful.

Teamwork is a crucial component in this activity. Selecting the process and solving the problem should be a shared experience.

If this activity is to be used as a hands-on assessment instrument, allow interaction during the problem solving and then instruct students to make a record of their work independently. The learning takes place in a group setting and the communication of the experience becomes a personal one and may provide a measure of what has been learned.

Management
1. Fill bags with small objects such as dried beans (baby limas, garbanzos, white navy beans), unpopped popcorn, small buttons, rice, small elbow macaroni, paper clips, small counters, etc. A suggested number range is 700-1500.
2. Begin with smaller numbers rather than larger numbers to ensure success. Keep objects uniform in size and shape. To increase the sophistication of the activity, increase number and vary the shapes and sizes in one bag.

Procedure
1. Ask the *Key Question* and state the *Learning Goals*.
2. Distribute a bag to each group of four to five students. Invite them to observe and describe the objects in their bags. Discuss the attributes that might affect the number of objects.

3. Have students record a wild guess as to the number of objects in the bag. Allow time for them to write the guesses of every person in their group, Tell them to find the average for the group.
4. Encourage students to work as a team to brainstorm a strategy to estimate more accurately the number of objects in the bag without counting each one. Discuss possible strategies such as measuring a part of the total and counting or determining mass to calculate number.
5. Have the students describe, step by step, the procedure in words and then in corresponding math symbols.
6. Direct them to apply the strategy and submit a revised estimate of number.
7. Give students time to count the actual objects in the bag. Have them record and compare their guess, their estimate, and the actual count.
8. Invite all groups to share their results. To determine the amount of error, find the difference between the estimate and the actual count for each group. If desired, you can indicate whether the amount of error represents an overestimate (+) or underestimate (−).
9. As a whole class, evaluate the plans for estimating.

Connecting Learning
1. How does the size or shape of the objects affect the number in the bag?
2. What was the range of guesses in each group?
3. How close was the average guess to each guess in the group?
4. Describe the strategy selected to estimate the number of objects in the bag.
5. How many different strategies are there for estimating the number of objects in the bag? Describe them.
6. What happens to our strategy when the size of the objects in the bag is not uniform? Suppose we have a bag of buttons of two or three different sizes.
7. What are you wondering now?

Sample Solution

Description in Words	Math Symbols
Use the balance to divide the whole set of beans into two equal parts. Each new part is one-half.	B = all the beans b/2 = 1/2 of the beans
Return one-half of the beans to the bag and use the balance to divide the remaining one-half into two equal parts.	1/2 divided by 2 = 1/4 of beans
Return one part (one-fourth) to the bag and again use the balance to divide the one-fourth into two equal parts now called one-eighth.	1/4 divided by 2 = 1/8 of beans
Return one-eighth of the beans to the bag and use the balance to divide the remaining one-eighth beans into two equal parts.	1/8 divided by 2 = 1/16 beans
Place one-sixteenth into the jar and count the remaining one-sixteenth. Multiply the number in one part (130) by the number of parts (16) to get the total number in the bag. Compare to the actual count. (actual = 2085)	16 x 130 = 2080 beans

* © Copyright 2010. National Governors Association Center for Best Practices and Council of Chief State School Officers. All rights reserved.

A Close Call

Key Question

How close can you estimate the number of objects in the bag?

Learning Goals

Students will:

- estimate the number of objects in a container, and

- devise a strategy for calculating a close approximation to the exact number without counting each one.

A Close Call

1. Describe the contents of your bag (size, shape, color, etc.)

2. My first guess is:

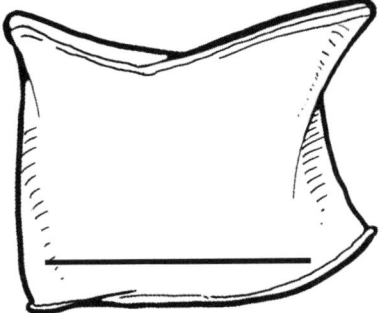

My team member guesses are:

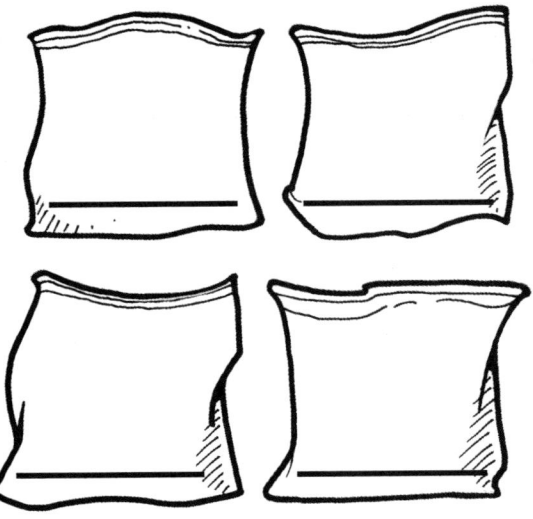

Their range is _____ to _____.

The average of our guesses is _____.

3. As a team, brainstorm a strategy to estimate the number in the bag without counting each one. Describe your strategy.

A Close Call

4. Describe, step by step, the procedure in words and in numerical symbols.

Words	Numbers and Symbols

Using our strategy, our team submits an estimate of _____ .
(number)

5. Compare the average guess, the estimate as a result of the strategy, and the actual count.

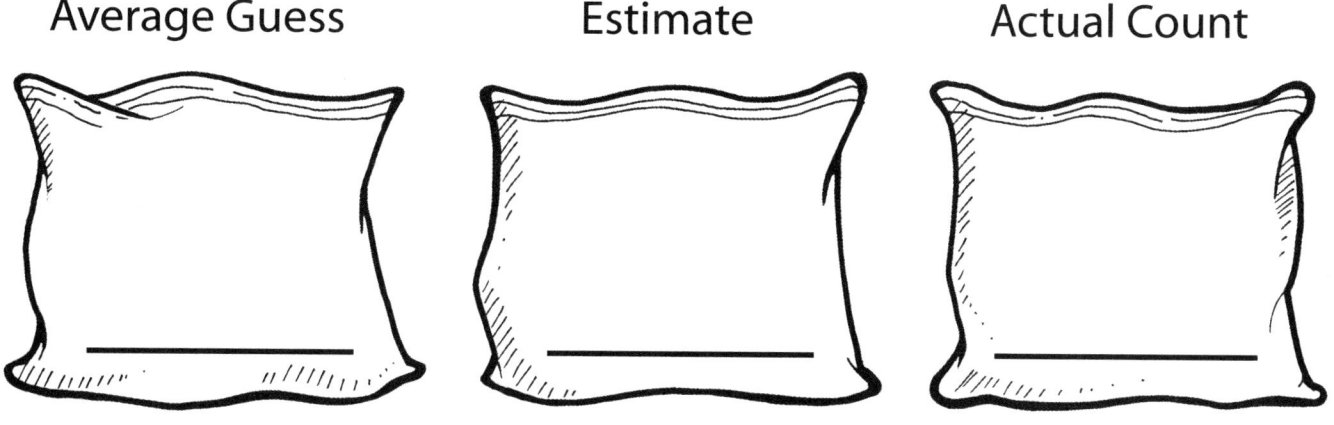

Average Guess   Estimate   Actual Count

A Close Call

6. Look at the class results.

Group	Objects	Average Guess	Estimate	Actual	Amount of Error
A					
B					
C					
D					
E					
F					

7. Evaluate the "good and bad" of a plan for estimating.

A Close Call

Connecting Learning

1. How does the size or shape of the objects affect the number in the bag?

2. What was the range of guesses in each group?

3. How close was the average guess to each guess in the group?

4. Describe the strategy selected to estimate the number of objects in the bag.

5. How many different strategies are there for estimating the number of objects in the bag? Describe them.

A Close Call

Connecting Learning

6. What happens to our strategy when the size of the objects in the bag is not uniform? Suppose we have a bag of buttons of two or three different sizes.

7. What are you wondering now?

Practice Problems

The problems on the following pages are provided for additional practice with the problem-solving strategies covered in this book. No strategies have been recommended for the individual problems, and they do not follow any particular order. Students must decide which strategy to use based on the individual problem. It is suggested that the problems be copied onto transparencies and cut apart. A problem can then be placed on the overhead as a "bright beginning" to start math class or at any time during the day when a few minutes are available for review. To receive maximum benefit from the problems, be sure to have a time of discussion after each one where the emphasis is on the process and strategies used rather than arriving at the correct answer.

How many blocks would you need to build a 3 x 3 x 3 cube?	Create a picture with pattern blocks that is $\frac{1}{3}$ yellow. Prove it.
If the hexagon is worth one, what is the value of each of the other pattern block pieces?	Find three equivalent fractions for $\frac{1}{4}$.
Mike is four years old. Andy is 14 years old. When will Andy be twice as old as Mike?	I have four coins. Two of my coins are the same. I have no nickels. The total value of my coins is 40¢. What coins do I have?
You have a bag of bears that contains four yellow bears and six red bears. What are the chances that you will pull out a yellow bear?	Tom has three times as many apples as Susan. Susan has one-fourth as many as Joe. Joe has four apples. How many do Tom and Susan have?

Colton's doctor told him to drink at least 40 ounces of milk each day. How many cups per day would that be? (1 cup = 8 ounces)	Boris ate eight pieces of pizza. Angela ate half as many pieces as Boris. How many pieces of pizza in all did they both eat?
Richard is cruising the highway at 65 miles per hour. How far will he travel in five hours?	Karen wants to buy 120 bottles of soda. There are 24 bottles in a case. How many cases does she need to buy?
How many capital letters have a vertical line of symmetry? ...horizontal line of symmetry? ... both?	You have $2.00 to buy marbles with. Aggies cost 14¢ each and boulders cost 18¢ each. How many of each kind can you buy for $2.00?
I have 33 cents in my pocket. I have nine coins and none of them are quarters. What coins do I have?	In the town parade, there are four rows with six boys in each row and nine rows with nine girls in each row. How many total students are there in the parade?

Jenny had $4.00. She bought four suckers that sold as two for 18¢, two candy bars for 65¢ each, and a notebook for $1.46. How much money did she spend?	You have three letters—A, E, and T. List all the three-letter combinations you can make using each letter once. How many of the combinations are words?
Using the digits 5, 6, 8, and 2, what is the largest odd number you can make? ...the largest even number? ...the smallest odd number? ...the smallest even number?	In Friday night's football game, South Hills lost to Central 20 to 12. What are the different ways South Hills could have scored the 12 points?
Karis and Ben were picking weeds for a neighbor. They picked 22 in the first five minutes on the job. Ben picked six more than Karis. How many did each pick?	Kate has not done very well in her astronomy class. Her test scores were 72, 81, 70, 79, 75, 84, and 97. What will her grade be on her progress report?

A rabbit is at the bottom of a pit that is 20 meters deep. Every day the rabbit climbs five meters up. Every night it slides three meters down. How many days does it take before the rabbit reaches the top of the pit?	Arrange the digits one to nine to form three numbers. Make the second number 111 more than the first number. Make the third number 711 more than the second number.
Carla and Noel were playing a factors game. Carla would name a factor of 24 and Noel had to give the other factor. Show the ways Carla and Noel could play the game.	Jennifer, Nanci, Scott, and Ebony were playing Monopoly. Jennifer won the game. Nanci placed ahead of Scott. Ebony placed ahead of Nanci. Who placed last?
Write the next three numbers in each sequence: 2, 4, 6, ___, ___, ___ 1, 4, 7, 10, ___, ___, ___ 1, 2, 4, 8, 16, ___, ___, ___ 1, 2, 4, 7, 11, ___, ___, ___	Write the next three numbers in each sequence: 1, 3, 5, ___, ___, ___ 21, 17, 13, ___, ___, ___ 0, 1, 3, 6, 10, ___, ___, ___ $\frac{1}{3}, \frac{2}{6}, \frac{3}{9}$, ___, ___, ___

At the store, three planes and a bird cage cost the same as five planes. If one plane costs $3.50, then what does a bird cage cost?	The total fare for two adults and three children on the *Puke and Hurl* ride is $14.00. If a child's fare is one half of an adult's fare, what is the adult fare?
Jo picked a number. She added 10 to it, doubled that sum, and then subtracted five. Her answer was 39. What number did Jo pick?	A group of kids was eating hot dogs. There were 18 hot dogs total. Each kid had an even number of hot dogs. No one had more than three hot dogs. How many kids were there?
One crab and one shark cost the same as four crabs. If one crab costs $6, then how much does one shark cost?	The calculator is showing 242 as the answer. What problems could have been put into the calculator to get this answer?

Campers had to choose one potato dish and one vegetable each night for dinner. Their potato choices were mashed, baked, or French fries. Their vegetable choices were corn or green beans. List all of the different dinners a camper could choose.

The Menendez family is planning their summer vacation. They can travel by plane, train, or car. They can go to Yosemite, Yellowstone, or the Grand Canyon. How many different trips could they take?

Shep the sheep dog is lazing around near the house. She counts the number of animals in the barnyard. There are 11 of them, all pigs and ducks. Then her eye runs over the legs. She sees 28 legs. How many ducks are there? How many pigs?

Make a 4 x 4 grid where the squares match these descriptions:
- four blue squares
- three red squares
- three white squares
- three green squares
- three yellow squares, and
- no color appears more than once in any row, column, or diagonal.

Carol's cat weighs two pounds. Doug's dog weighs 14 pounds. Both pets are gaining one pound a month. If they keep on gaining weight like that, the dog will soon weigh three times as much as the cat. How much will the cat weigh then?

I have six blocks glued together in a row on a table. I want to spray paint the blocks. Every block has six sides. The sides that touch the table or sides of other blocks don't get painted. How many sides of the blocks get painted?

Jamie went to her grandpa's farm. He has cows and chickens on his farm. She noticed that there were a total of 26 heads and 68 feet among them. How many chickens and how many cows does her grandpa have?

Priscilla Pig loves pink. Everything she has is pink. She just built a brick wall and she is going to paint it pink. The wall has 14 bricks across and is 11 bricks high. She is going to paint all sides of the bricks that are facing out. How many sides will Priscilla have to paint?

Brianna, Jason, and Trent each have some pens. One has six pens, one has 11 pens, and one has 17 pens. Figure out how many pens each person has. Jason does not have the most. If Trent gave Jason seven pens, Jason would have the most.

Put the yuk, ugh, bah, ick, and glok in order from biggest to smallest.
- A yuk is bigger than an ugh.
- A bah is not the smallest.
- A yuk is not the biggest.
- Only one thing is less than an ick.
- A glok is more than a bah.
- More than one thing is bigger than a bah.

There were 20 problems on a math test. Each correct answer earned five points. Two points were deducted for each wrong answer. Lenora got 72 points. How many correct answers did she have?

Ben, Letty, Tracey, and Dawn each ate a different kind of muffin for breakfast this morning. There were four different muffins on the table: apple, banana, lemon, and spice. Letty ate the lemon muffin. Dawn hates banana flavor. Ben doesn't like any fruit flavors. Which muffin did each eat?

Jessica has four bags of candy that she bought for $1.50 a bag. Each bag has six pieces of candy in it. How many more bags does she need to buy to give each of her 25 classmates one piece? How much will that cost?	Yesterday morning, there were more than five apples on the tree, but fewer than 11. This morning, the crows came and ate five apples. Now the number of apples left is more than four. How many apples are left on the tree?
Your plane leaves at 9:15. You have to arrive at the airport two hours before your scheduled departure time. It takes 45 minutes to drive to the airport. What time do you have to leave home?	In the morning, it takes you 10 minutes to shower, 15 minutes to get dressed and do your hair, and 20 minutes to eat and make a lunch. It takes you 20 minutes to walk to school. You get to school at 8:05. What time do you get up?

Lu went to Freddy's Funky Fashions. She spent half of her money on a dance outfit. She spent $100 of her remaining money on a pair of running shoes. Then she spent half of the money she had left on a purse. She had $40 left. How much did she have to start with?

Each weekend Lorenzo earns money delivering groceries. He deposits four-fifths of what he earns in his savings account and keeps the rest to spend. Last weekend, he kept $12. How much did Lorenzo earn last weekend?

A vendor sells two types of newspapers. One kind is 25¢ and the other is 40¢. One day she sold 100 papers and made exactly $28. How many of the 25¢ papers did she sell?

Jack walked from Santa Clara to Palo Alto. It took one hour and 25 minutes to walk from Santa Clara to Los Altos. Then it took 25 minutes to walk from Los Altos to Palo Alto. He arrived in Palo Alto at 2:45 P.M. At what time did he leave Santa Clara?

Yolanda and Will found that they were getting more and more snails in their garden. On the first day they counted nine snails, then on the second day there were 16. On the third day they counted 23. There were 30 on the fourth day, and 37 on the fifth day. If the pattern continues, on what day will they count more than 90 snails?

On her first day of vacation, Angela found two sand dollars on the beach. She put them in an old sock. The next day she found four sand dollars, and she put them in her sock. On each day of her vacation, Angela found two more sand dollars than she had found the day before. On what day did she have 42 sand dollars in the sock?

I had 500 jelly beans. On the first day of the month, I gave out 50 jelly beans. On the second day, I gave out 62 jelly beans—12 more than the first day. On the third day I gave out 74 jellybeans—12 more than the second day. If I gave out all the jellybeans at this rate, on what day of the month did I run out?

Ming folded her T-shirts and put them in two stacks in her drawer. She put three T-shirts in each stack. She put the green shirt under the blue shirt. She put the yellow shirt on the right side of the green shirt. She put the orange shirt on top of the blue shirt. Finally, Ming put the pink shirt between the yellow shirt and the red shirt. Where did Ming put each T-shirt in her drawer?

Joseph, Samantha, and Aree have the following jobs: teacher, chemist, and writer. Their salaries are $51,000, $46,600, and $41,400. Figure out the salary and job for each person.
- The writer earns $46,600.
- The teacher earns less than the chemist.
- Aree is not a teacher.
- Joseph is not a chemist.
- Samantha earns the most money.

Marty and Ramon want to ride the new *Freak Out* roller coaster at the fair. The boys can go into the fair through 3 different gates. They must stop at the main ticket booth that is just inside the gates, and they can take 4 different walkways from the ticket booth to the *Freak Out*. How many different paths can Marty and Ramon take from outside the fair to the *Freak Out*?

Mrs. Wong has misplaced her bank statement for March. Since there was little activity in her account that month, she is sure that she can figure out her March first balance. She knows that she withdrew one-third of her funds early in March, and later deposited checks for $150 on three separate days. She also remembers that her April first balance was $672. What was her March first balance?

Rachel's family consists of Kyle, Micah, Jennifer, and Abigail. They are Rachel's mother, father, younger brother, and younger sister. Match the names to the family members.
- Micah is older than Rachel.
- Abigail is not Rachel's younger brother.
- Jennifer is not Rachel's father. She is also not Rachel's younger sister.

Mr. and Mrs. Jeremy Mouse sell square tins of cheese buns in their shop. Right now they have 12 tins of buns in the shop. They stack the tins on the shelf, two tins in each stack, and the stacks touch sides. Every day Mr. Mouse dusts the sides of the tins that are not touching the shelf or another tin. How many sides must he dust on 12 tins of buns?

Alyssa, Cari, and Ethan each recycled a different number of cans (17, 25, and 28) and junk mail letters (121, 113, and 110). How many did each recycle?
- Alyssa recycled 85 more junk mail letters than the number of cans she recycled.
- Cari recycled less than 112 junk mail letters.
- Ethan recycled the most junk mail letters.
- Cari recycled the fewest cans.

There were 30 campers in Crazy Creek Park on Friday when it started to rain. In the first hour of rain, three campers took down their tents and left. In the second hour, six campers took down their tents and left. The rain kept pouring down. Every hour three more campers left than during the hour before. In what hour did the last campers leave the park?

At noon on Monday, 51 caterpillars met for lunch on the leaves of the old tree. On Tuesday, only 42 caterpillars came to the tree for lunch. On Wednesday, only 33 caterpillars came to the tree. The number of caterpillars coming for lunch kept decreasing in the same way each day. On what day did only six caterpillars come to the tree?

The AIMS Program

AIMS is the acronym for "**A**ctivities **I**ntegrating **M**athematics and **S**cience." Such integration enriches learning and makes it meaningful and holistic. AIMS began as a project of Fresno Pacific University to integrate the study of mathematics and science in grades K-9, but has since expanded to include language arts, social studies, and other disciplines.

AIMS is a continuing program of the non-profit AIMS Education Foundation. It had its inception in a National Science Foundation funded program whose purpose was to explore the effectiveness of integrating mathematics and science. The project directors, in cooperation with 80 elementary classroom teachers, devoted two years to a thorough field-testing of the results and implications of integration.

The approach met with such positive results that the decision was made to launch a program to create instructional materials incorporating this concept. Despite the fact that thoughtful educators have long recommended an integrative approach, very little appropriate material was available in 1981 when the project began. A series of writing projects ensued, and today the AIMS Education Foundation is committed to continuing the creation of new integrated activities on a permanent basis.

The AIMS program is funded through the sale of books, products, and professional-development workshops, and through proceeds from the Foundation's endowment. All net income from programs and products flows into a trust fund administered by the AIMS Education Foundation. Use of these funds is restricted to support of research, development, and publication of new materials. Writers donate all their rights to the Foundation to support its ongoing program. No royalties are paid to the writers.

The rationale for integration lies in the fact that science, mathematics, language arts, social studies, etc., are integrally interwoven in the real world, from which it follows that they should be similarly treated in the classroom where students are being prepared to live in that world. Teachers who use the AIMS program give enthusiastic endorsement to the effectiveness of this approach.

Science encompasses the art of questioning, investigating, hypothesizing, discovering, and communicating. Mathematics is a language that provides clarity, objectivity, and understanding. The language arts provide us with powerful tools of communication. Many of the major contemporary societal issues stem from advancements in science and must be studied in the context of the social sciences. Therefore, it is timely that all of us take seriously a more holistic method of educating our students. This goal motivates all who are associated with the AIMS Program. We invite you to join us in this effort.

Meaningful integration of knowledge is a major recommendation coming from the nation's professional science and mathematics associations. The American Association for the Advancement of Science in *Science for All Americans* strongly recommends the integration of mathematics, science, and technology. The National Council of Teachers of Mathematics places strong emphasis on applications of mathematics found in science investigations. AIMS is fully aligned with these recommendations.

Extensive field testing of AIMS investigations confirms these beneficial results:
1. Mathematics becomes more meaningful, hence more useful, when it is applied to situations that interest students.
2. The extent to which science is studied and understood is increased when mathematics and science are integrated.
3. There is improved quality of learning and retention, supporting the thesis that learning which is meaningful and relevant is more effective.
4. Motivation and involvement are increased dramatically as students investigate real-world situations and participate actively in the process.

We invite you to become part of this classroom teacher movement by using an integrated approach to learning and sharing any suggestions you may have. The AIMS Program welcomes you!

Get the Most From Your Hands-on Teaching

When you host an AIMS workshop for elementary and middle school educators, you will know your teachers are receiving effective, usable training they can apply in their classrooms immediately.

AIMS Workshops are Designed for Teachers
- Hands-on activities
- Correlated to your state standards
- Address key topic areas, including math content, science content, and process skills
- Provide practice of activity-based teaching
- Address classroom management issues and higher-order thinking skills
- Include $50 of materials for each participant
- Offer optional college (graduate-level) credits

AIMS Workshops Fit District/Administrative Needs
- Flexible scheduling and grade-span options
- Customized workshops meet specific schedule, topic, state standards, and grade-span needs
- Sustained staff development can be scheduled throughout the school year
- Eligible for funding under the Title I and Title II sections of No Child Left Behind
- Affordable professional development—consecutive-day workshops offer considerable savings

Call us to explore an AIMS workshop
1.888.733.2467

Online and Correspondence Courses
AIMS offers online and correspondence courses on many of our books through a partnership with Fresno Pacific University.
- Study at your own pace and schedule
- Earn graduate-level college credits

See all that AIMS has to offer—visit us online

 http://www.aimsedu.org

Check out our website where you can:
- preview and purchase AIMS books and individual activities;
- learn about State-Specific Science and Essential Math;
- explore professional development workshops and online learning opportunities;
- buy manipulatives and other classroom resources; and
- download free resources including articles, puzzles, and sample AIMS activities.

 find us on facebook

Become a fan of AIMS!
- Be the first to hear of new products and programs.
- Get links to videos on using specific AIMS lessons.
- Join the conversation—share how you and your students are using AIMS.

While visiting the AIMS website, sign up for our FREE *AIMS for You* e-mail newsletter to get free activities, puzzles, and subscriber-only specials delivered to your inbox monthly.

SOLVE IT! 4th © 2012 AIMS Education Foundation

AIMS Program Publications

Actions With Fractions, 4-9
The Amazing Circle, 4-9
Awesome Addition and Super Subtraction, 2-3
Bats Incredible! 2-4
Brick Layers II, 4-9
The Budding Botanist, 3-6
Chemistry Matters, 5-7
Concerning Critters: Adaptations &
    Interdependence, 3-5
Concerning Critters: Observations &
    Classification, 3-5
Counting on Coins, K-2
Cycles of Knowing and Growing, 1-3
Crazy About Cotton, 3-7
Earth Book, 6-9
Earth Explorations, 2-3
Earth, Moon, and Sun, 3-5
Earth Rocks! 4-5
Electrical Connections, 4-9
Energy Explorations: Sound, Light, and Heat, 3-5
Exploring Environments, K-6
Fabulous Fractions, 3-6
Fall Into Math and Science*, K-1
Field Detectives, 3-6
Floaters and Sinkers, 5-9
From Head to Toe, 5-9
Getting Into Geometry, K-1
Glide Into Winter With Math and Science*, K-1
Gravity Rules! 5-12
Hardhatting in a Geo-World, 3-5
Historical Connections in Mathematics, Vol. I, 5-9
Historical Connections in Mathematics, Vol. II, 5-9
Historical Connections in Mathematics, Vol. III, 5-9
It's About Time, K-2
It Must Be A Bird, Pre-K-2
Jaw Breakers and Heart Thumpers, 3-5
Looking at Geometry, 6-9
Looking at Lines, 6-9
Machine Shop, 5-9
Magnificent Microworld Adventures, 6-9
Marvelous Multiplication and Dazzling Division, 4-5
Math + Science, A Solution, 5-9
Mathematicians are People, Too
Mathematicians are People, Too, Vol. II
Mostly Magnets, 3-6
Movie Math Mania, 6-9
Multiplication the Algebra Way, 6-8
Out of This World, 4-8
Paper Square Geometry:
    The Mathematics of Origami, 5-12
Popping With Power, 3-5
Positive vs. Negative, 6-9
Primarily Bears*, K-6
Primarily Critters, K-2

Primarily Magnets, K-2
Primarily Physics: Investigations in Sound, Light,
    and Heat Energy, K-2
Primarily Plants, K-3
Primarily Weather, K-3
Probing Space, 3-5
Problem Solving: Just for the Fun of It! 4-9
Problem Solving: Just for the Fun of It! Book Two, 4-9
Proportional Reasoning, 6-9
Puzzle Play, 4-8
Ray's Reflections, 4-8
Sensational Springtime, K-2
Sense-able Science, K-1
Shapes, Solids, and More: Concepts in Geometry, 2-3
Simply Machines, 3-5
The Sky's the Limit, 5-9
Soap Films and Bubbles, 4-9
Solve It! K-1: Problem-Solving Strategies, K-1
Solve It! 2nd: Problem-Solving Strategies, 2
Solve It! 3rd: Problem-Solving Strategies, 3
Solve It! 4th: Problem-Solving Strategies, 4
Solve It! 5th: Problem-Solving Strategies, 5
Solving Equations: A Conceptual Approach, 6-9
Spatial Visualization, 4-9
Spills and Ripples, 5-12
Spring Into Math and Science*, K-1
Statistics and Probability, 6-9
Through the Eyes of the Explorers, 5-9
Under Construction, K-2
Water, Precious Water, 4-6
Weather Sense: Temperature, Air Pressure, and
    Wind, 4-5
Weather Sense: Moisture, 4-5
What on Earth? K-1
What's Next, Volume 1, 4-12
What's Next, Volume 2, 4-12
What's Next, Volume 3, 4-12
Winter Wonders, K-2

Essential Math
Area Formulas for Parallelograms, Triangles, and
    Trapezoids, 6-8
Circumference and Area of Circles, 5-7
Effects of Changing Lengths, 6-8
Measurement of Prisms, Pyramids, Cylinders, and
    Cones, 6-8
Measurement of Rectangular Solids, 5-7
Perimeter and Area of Rectangles, 4-6
The Pythagorean Relationship, 6-8
Solving Equations by Working Backwards, 7

* Spanish supplements are available for these books. They are only available as downloads from the AIMS website. The supplements contain only the student pages in Spanish; you will need the English version of the book for the teacher's text.

For further information, contact:
AIMS Education Foundation • 1595 S. Chestnut Ave. • Fresno, California 93702
www.aimsedu.org • 559.255.6396 (fax) • 888.733.2467 (toll free)

Duplication Rights

No part of any AIMS books, magazines, activities, or content—digital or otherwise—may be reproduced or transmitted in any form or by any means except as noted below.

Standard Duplication Rights

- A person or school purchasing AIMS activities (in books, magazines, or in digital form) is hereby granted permission to make up to 200 copies of any portion of those activities, provided these copies will be used for educational purposes and only at one school site.
- For a workshop or conference session, presenters may make one copy of any portion of a purchased activity for each participant, with a limit of five activities or up to one-third of a book, whichever is less.
- All copies must bear the AIMS Education Foundation copyright information.
- Modifications to AIMS pages (e.g., separating page elements for use on an interactive white board) are permitted only within the classroom or school for which they were purchased, or by presenters at conferences or workshops. Interactive white board files may not be uploaded to any third-party website or otherwise distributed. AIMS artwork and content may not be used on non-AIMS materials.

Standard duplication rights apply to activities received at workshops, free sample activities provided by AIMS, and activities received by conference participants.

Unlimited Duplication Rights

Unlimited duplication rights may be purchased in cases where AIMS users wish to:
- make more than 200 copies of a book/magazine/activity,
- use a book/magazine/activity at more than one school site, or
- make an activity available on the Internet (see below).

These rights permit unlimited duplication of purchased books, magazines, and/or activities (including revisions) for use at a given school site.

Activities received at workshops are eligible for upgrade from standard to unlimited duplication rights.

Free sample activities and activities received as a conference participant are not eligible for upgrade from standard to unlimited duplication rights.

State-Specific Science modules are licensed to one classroom/one teacher and are therefore not eligible for upgrade from standard to unlimited duplication rights.

Upgrade Fees

The fees for upgrading from standard to unlimited duplication rights are as follows.
For individual activities, the cost is $5 per activity per school site.
For Literature Links bundles, the cost is $12 per bundle per school site.
For books, the cost is based on the price of the book (see table).

Book Price	Upgrade Fee
$9.95	$15.00/site
$18.95	$24.00/site
$21.95	$27.00/site
$24.95	$30.00/site
$29.95	$35.00/site
$34.95	$40.00/site
$49.95	$55.00/site

The cost of upgrading is shown in the following examples:
For five activities at six schools:
    5 activities x $5 x 6 schools = $150

For two books (at $21.95) at 10 schools:
    2 books x $27 x 10 schools = $540

For three books (at $24.95) and four activities at eight schools:
    (3 books x $30 x 8 schools) + (4 activites x $5 x 8 schools) = $720 + $160 = $880

Purchasing Unlimited Duplication Rights

To purchase unlimited duplication rights, please provide us the following:
1. The name of the individual responsible for coordinating the purchase of duplication rights.
2. The title of each book, activity, and/or magazine issue to be covered.
3. The number of school sites and name and address of each site for which rights are being purchased.
4. Payment (check, purchase order, credit card).

Requested duplication rights are automatically authorized with payment. The individual responsible for coordinating the purchase of duplication rights will be sent a certificate verifying the purchase.

Internet Use

AIMS materials may be made available on the Internet if all of the following stipulations are met:
1. The materials to be put online are purchased as PDF files from AIMS (i.e., no scanned copies).
2. Unlimited duplication rights are purchased for all materials to be put online for each school at which they will be used. (See above.)
3. The materials are made available via a secure, password-protected system that can only be accessed by employees at schools for which duplication rights have been purchased.

AIMS materials may not be made available on any publicly accessible Internet site.